Living With Your Heart & Lungs

By: Stanley L. Bryant, CRT, EMT-P, AS-CIS, BSCS, MA
Glenn Obst, CRT

Living With Your Heart and Lungs

ISBN 978-0-615-24001-5

Living With Your Heart & Lungs

By: Stanley L. Bryant, CRT, EMT-P, AS-CIS, BSCS, MA
Glenn Obst, CRT

Introduction

In the event of a disaster or even as mom and dad age naturally, are you prepared to assist them with medical decisions, do you know what to do to keep them safe, would you know what to do to help them on a daily basis? Medical facilities are all ready overloaded and would become nonfunctional after a disaster. Everyone needs the easy to read information contained in this book to help our loved ones and family members.

Baby Boomers are becoming Geri Boomers (older patients) the growth of our older population is, for the lack of words, booming. Older adults have unique needs and require more attention than most people in the younger generations. Today everyone needs information and a support group of informed people they can trust when making complicated decisions of how best to manage their medical care. Mom, dad, grandma and grandpa often understand far less than we give them credit and remember even less when it comes to medical care. It will at some point, be up to you to bear the responsibility of understanding and guiding your loved ones in the decision-making process. This book is your vital resource of trusted information, conveniently presented in words that you and your loved ones can understand in order to make informed medical decisions.

The field of cardiopulmonary medicine (heart and lungs) becomes more complex with every pandemic, each new disease (like Swine Flu, H1N1 which has now mutated to H1N3 etc.) and each medical breakthrough. In addition, as people age the body changes which affects how medications work, how diseases present, how we view illness and how we communicate effectively. Understanding the illness, tests, treatments and available treatment options has become staggering. With the imminent changes to our healthcare system, it is paramount that patients today become informed medical consumers (medical advocates). You, your family and loved ones will have limited financial and reimbursement resources to spend on medical care now and in the future. This is evidenced by the pharmaceutical commercials marketing their medications to you on television and radio each day "ask you doctor if our medication is right for you". The lack of control by the courts in medical malpractice law suits often forces your healthcare provider to practice defensive medicine. This has made all medical care tremendously expensive due to the extensive and unnecessary testing required to help protect the healthcare provider from lawsuits.

In the days gone by, back when life was much simpler, doctors and nurses were "wonder-workers" held in the highest esteem and never questioned. Today, you must be able to communicate problems and discuss what works for you, your concerns and your goals during the very limited appointment time with your healthcare provider. Over 75% of the doctors in the United States can not accept new patients, the practices are just full. This book is an easy to use, comprehensive and an informative reference bridging the gap of your medical needs and healthcare literacy (**it is written in words you can understand**). It contains a multitude of relevant information intended to be helpful in communicating with your healthcare professional and in making health and wellness decisions for a loved one or for yourself. All of the clinical information contained in this book has been compiled from numerous trusted medical reference sources.

Acknowledgements:

This book is dedicated to Bryan Obst, Barry Swan, Orlo Elliott (Cindy's dad), to our loving wives Wendy and Regina, and to all of those patients that did not have a book such as this, to help to guide them and their families with decisions in this most complicated field of medicine. It is simply amazing that this information has not been compiled sooner.

Forward

This book has been written in tribute to those brave people who are suffering from any number of debilitating diseases or conditions. Whether it is COPD (Chronic Obstructive Pulmonary Diseases) or CHF (Congestive Heart Failure), there are no "silver bullet" cures. These diseases are life altering, both physically and mentally. The inability to engage in one's favorite activities, or in more dramatic cases, the loss of the ability to walk to the mailbox, put on make up, shave, or take a shower without becoming short of breath is depressing and demoralizing. As Respiratory Therapists, we understand how debilitating heart and lung (cardiopulmonary) diseases become and how they dramatically affect one's outlook and attitude. We offer this book to you and your loved ones with the hope that it might afford some illumination to a dark and confusing health situation. It takes great courage and mental fortitude to battle these unrelenting diseases which took years to develop. Our wish for you is that you will be able to keep a positive attitude and take pride in the fact that you are doing all that you can to preserve your health and well-being, for yourself and for your loved ones.

A Note from the Authors

The first point we should make is that this is not a textbook and is not designed to be read from cover to cover. In the first six chapters you will find the information that is critical to your healthcare and safety. In the last six chapters you will find reference materials (what we call the good stuff) so you can read in normal English what the doctor is talking about, testing for and why he/she is prescribing medications.

Second, this book is yours, make notes in it, highlight it, take it to your appointments with your healthcare providers and share it with as many friends as you can. If a friend or loved one could benefit from some of its pages, copy them and share it with them. It is designed to help you to understand, maintain the best quality of health possible and money.

Third, please remember this book is not designed to diagnose or treat any disease processes, but it will help you to understand what your doctor is doing and what you can do to work as a member of your healthcare team effectively.

References

1. Wilkins, Robert L., Scanlan, Craig L., Stoller, James K.: Egan's Fundamentals of Respiratory Care, 8th edition. 2003. Elsevier Science.

2. Merck Manual, Robert Berkow MD, Andrew Fletcher, MB, 18th edition, 2006

3. Tabor's Cyclopedic Medical Dictionary, Clayton Thomas, MD, 21st edition, 2009

4. Murray, John F., Nadel, Jay A.: Textbook of Respiratory Medicine, 3rd edition. 2000. WB Saunders Company.

5. Atlas of Pathophysiology, Judith Schilling MaCann, RN, MSN, Lippincott, 2nd edition, 2006

6. Giddens, Jean Foret, PhD, MSN, RN, CS, Langford, Rae W., EdD, RN: Nursing PDQ. 2004. St Louis, Mosby.

7. Corning, Helen Schaar, RRT, Bryant, Stanley L. CRT: Mosby's Respiratory Care PDQ. 2006. Elsevier Science.

8. LaSage, Paul EMT-P, Assistant Chief, Derr, Paula, RN, BSN, CCRN, CEN, Tardiff, Jon, Paramedic, EMS Field Guide ALS Version, 13th Edition 2001. InforMed.

9. WebMD Health, on-line, www.webmd.com. Accessed July, 2009.

10. Oakes, Clinical Practitioner's Pocket Guide to Respiratory Care, Oaks, Dana 6th Ed 2004.

11. Gold, Warren M., Nadel, Jay A.: Atlas of Procedures in Respiratory Medicine. 2002. WB Saunders Company.

12. http://www.nasa.gov, 2009

13. Surgeon General's 2000 report on Smoking.

14. American College of Critical Café Medicine review, February 2007.

15. American Lung Association 1-800-330-5864.

16. American Cancer Society 1-800-ACS-2345

17. Scott Sand CEO & Chairman Ingen Technologies, Inc. www.ingen-tech.com March 2007

18. Quitting Smoking for Dummies, Wiley Publishing, Inc., 2004

19. http://medlineplus.gov/, 2010

20. http://www.nih.gov/, 2010

21. http://www.fda.gov/, 2010

22. http://www/nlm.nih.gov, 2010

23. www.ama-assn.org, 2010

24. http://www.webmd.com, 2010

25. National Sleep Foundation website http://www.sleepfoundation.org, 2009, 2010

26. University of Pittsburg Emphysema Research Center, Director Frank Sciurba, MD. Feb 2008

27. Delmar's Respiratory Care Drug Reference, Fred Hill, Cengage Delmar Learning, Jan 99

28. Government Regulations http://www.regulations.gov, 2009, 2010

29. Drug Information Online http://www.drugs.com, 2010

30. World Wide Metric Conversions http://worldwidemetric.com/metcal.htm, 2008, 2009

31. World Health Organization, http://www.who.gov, 2010

32. The United States Department of Health and Human Services, writing/literacy recommendations for medical reading 2009.

33. American Academy of Sleep Medicine, http://www.aasmnet.or/resources/clinical guidelines, 2009

34. American Association of Retired People, http://bulletin.aarp.org/, 2009

35. National Institute of Health, Medline Plus, http://www.nlm.nih.gov/medlineplus/ency, 2010

36. Pulmonary Education and Research Foundation, http://www.perf2ndwind.org, 2010

37. American Heart Association http://www.anericanheart.org, 2010

Table of Contents

Abbreviations

ABG	Arterial Blood Gas		MDI	metered dose inhaler
ACE	Angiotensin-converting enzyme		mEq	milliequivalent
asa	Aspirin		mg	milligram
BID	two times daily		min	minute
BiPAP	Bi-level Positive Airway Pressure		ml	milliliter
CHF	congestive heart failure		mm	millimeter
cm	centimeters		mmHg	millimeters of mercury or Torr
CNS	central nervous system		ml	milliliter
CPAP	Continuous Positive Airway Pressure		mm	millimeter
CPT	Chest Physical Therapy		neb	nebulizer
dl	deciliter		NSAID	Non-steroidal anti-inflammatory drug
DPI	dry powder inhaler		oz	ounces
Dx	Diagnosis		PO	by mouth
EKG	Electrocardiogram		Prn	as needed
f	frequency or rate		psi	pounds per square inch
FiO2	Amount of inspired oxygen		pt	patient
g	gram		Q	every
G+	gram positive		QD	once every day
G–	gram negative		QHS	Bedtime (hours of sleep)
gtt	drop		QID	four times daily
HCTZ	Hydrochlorothiazide		SL	sublingual
HTN	Hypertension		SQ	subcutaneous
hr	hour		Tab	tablet
Hx	History		TID	three times daily
IM	intramuscularly		Tx	treatment
IO	Intraosseous (in the bone)		μ	micro
IU	International Unit		μg	microgram
IV	Intravenous		U	units
kg	kilogram		vent	ventilator
L	liter		vol	volume
lpm	Liters per minute		w/a	while awake
lb	pound		w/o	without
PFT	Pulmonary Function Test		WOB	work of breathing
mcg	microgram		Z	No meaning, just like the letter

We have used very few abbreviations in this book. We have listed abbreviations here and within the text for your convenience and reference. Most abbreviations will soon be outdated due to the possibility of mistaking the meaning and causing medication and treatment errors. Most healthcare practitioners are required to use plane language when writing in your medical chart to reduce the chance of medical errors.

Chapter 1
What I Should Know

People learn by doing, seeing, reading and/or listening, hence after you read any instructions try setting up the device, or delivering the medication for yourself. If you have questions or problems, your homecare provider (particularly your Respiratory Therapist), local rehabilitation center (a terrible name), healthcare provider or the Internet can help you.

Each year our healthcare system wastes trillions of dollars on healthcare because people do not understand their own role in healthcare. People just do not care or ignore their role in healthcare until a family member or loved one is seriously ill and the healthcare system no longer can afford this attitude. Everyone must understand that their are four major causes of death in America, heart disease, cancer, cerebrovascular (bleeding or blockages in the head) disease and COPD (Chronic Obstructive Pulmonary Disease). These diseases have four central causes and are in large part preventable. The primary causes of disease include high fat diets, excessive alcohol use, use of tobacco products and inactivity. You alone control your destiny, and quality of life. Historically, healthcare providers have been trained to fix what is wrong and not to prevent it from happening in the first place. This attitude is changing, but change takes time. If you want to live longer, you must take charge of your health now. You must learn more about your body, become informed enough to ask the right questions (will be covered in detail in Chapter 7), and be willing to discuss treatment options with your healthcare provider. Do not forget that a positive attitude is a crucial part of healing.

Learn the facts, you are not the only one involved in your healthcare, your family has a stake in your health also. The single most important thing you can do to improve you health is to take part in your care. Get you annual immunizations, blood tests and preventative screenings done to protect your health.

Know Your Body

When you notice anything that is out of the ordinary for you, make a note about it with the time and date for your healthcare provider. Your notes on seemingly unimportant signs or symptoms will help your healthcare provider better care fro you.

Here is an example: Pitting Edema on your arms or legs. Sounds like a big term, but it is not really. All it means is you have excess fluid or fat build up under your skin and when you firmly press your finger on to the area the indentation (or pit made by your finger tip) does not go away for a longer period of time than normal. It could be only fat build up or it could mean that you are retaining water (kidneys are not working well or that your heart can not overcome gravity and pull the fluids back up and you may need a diuretic (water pill) prescribed. It often helps to elevate your feet (really elevate them, above the level of your heart) in order to allow the swelling to go back down.

Take Notes and Ask Questions

Take notes in a medical diary when events happen so your healthcare provider can better address the problem. Symptoms are an indication of the problem and are important. It may help you to keep a medical diary or calendar of events of what you have been doing to solve a medical or physical issue. By keeping a record, your healthcare provider will be able to see what you are doing on a daily basis.

Talk to your healthcare provider and pharmacist. Confide in them about your health questions and concerns. Do not assume that they are "all knowing" just because of their professions, but be assured that they will know where to find the answers to your questions (if it is not one they are asked quite often). If the truth were known, they will be pleasantly surprised and very much impressed with your active interest in your healthcare. Therefore, the bottom line is, talk to your doctor and pharmacist. Be an active participant, make notes in this book and take it with you to the appointment.

Research your diseases, the care and medications you are prescribed. You will discover that the number of sources and the sheer volume of information is overwhelming (i.e. doctors, hospitals, rehabilitation centers, specialty centers, internet, books etc.). With this in mind, the information must be able to be trusted and presented in such a way that you are able to understand. When doing research for this book we discovered many sources of outdated information and internet sites where the information is just wrong or misleading. Remember, there are no silver bullets for most healthcare problems.

Be an Informed Medical Consumer (Medical Advocate)

Do not base your healthcare provider selection on anything other than quality of service. If you are not satisfied, uncomfortable or not receiving the help and information you need, it is time to find a new healthcare provider and the support you deserve. Do not forget, titles do not make anyone an expert or the right one for you. You are paying for a service, you have worked hard and long to ensure you have the money and time to do the things you have planned.

Your healthcare team, hospitals and specialty care facilities constructed of people just as any other team and you are a member of that team. You have the responsibility to be an informed medical consumer (medical advocate) and an active member of your team. Ask the right questions, take notes and research your healthcare plan in order to make recommendations in your healthcare program. You will hear it repeatedly, "you are your last line of defense".

Do your homework and get your questions answered. If you not 100% sure of their answers or things do not feel right consider a second opinion or a different facility. Always be skeptical when it comes to your healthcare (just as you would about your money).

Literacy (Understanding what you read)

Literacy is the most significant pitfall for all people when it comes to healthcare. Unless you are a medical professional, you probably do not understand or remember much your healthcare provider tells you. You need to ask your healthcare provider to explain all of your questions in words you can understand and you need to take notes. Taking notes seems silly however, after you leave the doctor's office or even after they leave the exam room, you will forget most of what they said. The key to being a member of your healthcare team boils down to being able to do your research and understanding what you have read.

2

Doctors and Healthcare Providers

Doctors and healthcare professionals are people too. Communications with your healthcare provider is your responsibility. It can be difficult and sometimes a little scary. Just remember they are no smarter than you are, they just has more knowledge of medicine due to their extensive training.

Rules for all of your Healthcare Providers

Your healthcare provider must listen to you because you know your body better than anyone else could.

If he/she cannot explain your condition or the procedure you need, you need to find a new provider. Never accept answers like "it's too complicated to explain" or "you would not understand the explanation".

Your healthcare provider should always phone, email or mail you the test results as soon as possible, good or bad. There is nothing worse than setting around worrying about those test results and imagining the worst. Just remember, there are times when your test results should be discussed face to face with your healthcare provider.

Doctor Appointment

The best time for your doctor's appointment is early on Monday morning. The staff is rested and ready to go and emergencies have not pushed appointments times yet.

You doctor and his/her staff has an allotted time for your appointment and your healthcare needs. He/she must spend enough time in the examination room to solve your concerns. You too must understand that when you make an appointment when you are sick the time allotted will be just long enough to solve that problem. When you make you annual medical appointment for your check up he/she will allow you more time for the appointment in order to address your concerns.

Bring your notes and questions with you, with the understanding that the doctor and his/her staff may have to get back to you on some of your answers depending on the time available. If you have more than one problem to discuss, you must be prepared to make additional appointments to resolve the issues. Doctors are trying to help you and to make a living (they do not make near as much as you might think) so, work with them as a member of the team.

Note: Depending of how many problems you have, you may be required to make additional appointments to discuss your issues.

Note: If you are not going to make it to your appointment, call and cancel.

Diagnostic Testing and Treatment

Your healthcare provider will utilize tests, evaluations, symptoms, your medical history, what you tell them and their experience to diagnose your illness. Many diseases or illnesses have similarities and how you feel is an important sign or symptom. You need to communicate how you feel in order for your healthcare provider to accurately diagnose and treat your issue. Do not be a bystander in your healthcare or be afraid to get your healthcare providers attention. Chapter 9 will help you better understand an illness but, never try to self diagnose your symptoms (your healthcare providers have had many years of training).

You may think you have asthma based on your symptoms. However, did you know CPOD, allergies, gastric reflux, Cystic Fibrosis, genetics, post-nasal drip and many other processes can mimic asthma. Let your healthcare provider diagnose and treat your problems. If you need help with a problem, tell them.

Your healthcare provider would be happy to help you with any problems that may arise, but you must make them aware of them. You may have more than one problem to discuss, just be aware additional problems may require additional appointments, but if you feel it is important, let your healthcare provider decide about waiting for another appointment.

When you talk with your healthcare providers always tell them the truth.

If you lie to them about symptoms or problems, you cannot expect to be effectively treated.

Do not be too embarrassed to discuss your concerns. Many disease processes can be slowed or cured, but to receive the proper care, you have to be forthright (tell the truth) with your healthcare provider.

Now knowing all this, you must also understand that many healthcare providers practice defensive medical management when ordering tests and medications. Almost 1/3 of all tests and medications they prescribe are for their legal protection. Many others prescribed are only to treat symptoms rather than what is causing the problems. These unnecessary tests and medications greatly add to your medical care co pays and your overall cost of medical care. Ask your healthcare provider if the test or medication is necessary before you agree to get them done or filling the prescriptions.

Note: Never stop taking your medications or cancel a diagnostic test before you discuss it with you healthcare provider.

When your test results are sent to your healthcare provider remember, people are different when it comes to the results of medical testing. Normal for you may not be normal for others and vise versa. So don't be ignore or worry about any test result until your speak you your healthcare provider.

Medications

The FDA (Food and Drug Administration), pharmaceutical companies and pharmacies take great care in protecting the public when using prescribed medications. However, as always the final defense is your responsibility. You must be an informed consumer and never take any medications or herbs blindly. Remember, the FDA (Food and Drug Administration) does not routinely test or regulate over the counter herbs and supplements (about 70,000 different ones). Once identified as a problem or health hazard, the FDA will decide how the herb or supplement is controlled. There is no guarantees of potency (too much or too little), that they are impurity free (contaminates could be inside, i.e. lead, arsenic etc.) and that they are free from added chemicals or medications (i.e. caffeine, tranquilizers, amphetamines etc.) The U.S. Pharmacopeia (USP) is a voluntary program offered to manufactures that will certify the quality, purity and potency of dietary supplements (it would be wise to look for the USP stamp). You must understand the pros and cons of each medication you take because most medications (even herbs and supplements) have side effects or long term risks. You as a consumer must understand why you are taking your medications and what could possibly go wrong down the road. You and your healthcare provider can discuss these issues and decide what is best in your particular situation.

Never share your medications or take medications not prescribed by your healthcare provider and never buy medications over the internet.

As we age, we tend to forget things. Do not be too stubborn or proud to let our loved ones check on our medication supplies and help with reordering or setting up your daily medication dispensers. It never hurts to have a second pair of eyes when it comes to our health.

If you are not taking your medications as prescribed, tell them why.

> Some patients decide they do not need a medication, or they need less, or even more of a medication. This is dangerous and makes it quite difficult for your healthcare provider to make the proper changes in your dosage. Talk to your healthcare provider about what works and what does not. Many alternatives are available and you are responsible for communicating to your healthcare provider how your medical care is working.

If you cannot afford your medications, tell your doctor.

> There are many programs available to assist you with the costs of medications; some are pharmaceutical company programs, some are state programs, some are local, and sometimes you healthcare provider can offer free samples. Just ask…

If you have taken someone else's medications, own up.

> Prescribed medications are complicated and can be dangerous. Never take a medication a friend offers you. Someone might say to you, "it's a wonder pill, it sure helped me, I feel like new, try it if it works ask you doctor for some". Many medications will react with others and can be life threatening. Medications come in various strengths (i.e. 3mg, 1mcg, 1g etc) and if you run out of your medicine, someone else's may not be the same. As an endnote, if you hear about medication (i.e. TV, radio, friend, etc.) discuss them with your healthcare provider.

If you use herbs, your healthcare provider and pharmacist need to know.

> Some herbs can interact with prescribed medications. If you are taking herbs, make sure you tell your healthcare provider and pharmacist. Do not stop taking the herbs without discussing it with your healthcare provider. Keep yourself save and "just tell them".

Note: Just because it is over the counter, it does not mean it is safe or without side effects. In addition, many over the counter medications contain the same medications which can make an overdose possible. Read the label and talk to the pharmacist, it may just save your life.

Black Box Warning (also called black warnings label or boxed warning)

These Food and Drug Administration (FDA) warnings appear on the package insert for prescription medications (that is that big folded up paper with writing only a ten year old could read without a magnifying glass) that have been identified that may cause serous adverse effects or be life-threatening conditions. It is important to understand you are not taking prescription medications because you like the taste or just have extra money to spend. You and your healthcare provider have discussed the risks and decided that the benefits out way the risks. In other words, you stay alive now and worry about any adverse side effects later. You are your own last line of defense. When you are given a new prescription or medications you have never taken before, you must be an informed medical consumer (medical advocate) and ask questions. You should talk to your healthcare provider and pharmacist. Always make sure you understand the medication's benefits and its risks.

Medication, Food and Beverage Interactions

Medication interactions occur when two of more drugs you swallow, inhale, inject or wear on you skin react together changing the intended effect of your medications. Drug interactions may make your medications less effective, cause side effects or increase the actions of your medications. You are the last line of defense when it comes to drug interactions that can be harmful to your health so it is important that you are an informed medical consumer (medical advocate).

We have discusses how over the counter medications, vitamins and herbs can affect how your prescribed medications function but, few people know foods can do the same thing. As we age we are often prescribed more medications (sometimes a lot more) and understanding food interactions with these medications can save you money, save your healthcare provider time and may just save your life.

There are thousands of food and medications interactions however; we will only discuss a few common ones. Your doctor and pharmacist would be happy to discuss foods that interact with the medications you are taking. You need to be aware of these issues and ensure your healthcare team is aware of any allergies you have. Here are the basics:

Alcohol	Alcohol interacts with almost every medication and can cause liver damage which is where medications are cleared from your blood stream (many prescription and non-prescription medications are destructive to your liver). Alcohol can increase the effects of antihistamines, narcotics (analgesics or pain medications) and sedatives (sleeping pills) placing you at risk of an overdose. In addition, it can interact with antidepressant, bronchodilators and any medication that affect the brain or nervous system.
Candy	We know that blood sugar is elevated when your eat candy however; licorice (made from plant roots) interacts with many prescription and over the counter medications. Licorice can

6

interact with cardiac medications, or cause your potassium level to drop when taking water pills (diuretics). It can interfere with depression medications (i.e. MAOI's or Monoamine Oxidase Inhibitors) and increase the effects of oral corticosteroids (steroids). In addition, chewing tobacco and eating licorice can cause an unbalance of electrolytes in your blood (i.e. sodium, potassium etc.).

Chocolate   Chocolate contains caffeine and tyramine (see tyramine below) when combined with bronchodilators (beta agonists like Alabuterol) can make you shake or cause severe anxiety. Consuming large quantities can cause drug toxicity (can be poisonous).

Citrus fruits   Grapefruit or grapefruit juice can cause many medications (i.e. high blood pressure medications, some cardiac medications and medications taken for high cholesterol) to enter the bloodstream faster than normal or enhance their effects. They can also reduce your body's absorption ability of some antihistamines and decrease the effectiveness of antibiotics.

Note: Grapefruit and grapefruit juice affect almost 60 prescription medications.

Coffee, tea and pop   Caffeine when combined with bronchodilators (or beta agonists like Albuterol) and/or stimulants can make you shake or become anxious. Using caffeine with sleeping pills, well, that makes no sense at all. It can also reduce the effect of antibiotics. Consuming large quantities can cause drug toxicity (can be poisonous).

Egg products   Most seasonal flu vaccines are grown in eggs so if you are allergic to egg products you should discuss receiving flu vaccinations.

Garlic   Garlic can cause a dangerous decrease in blood sugar and overly enhance the effect of blood thinners (anticoagulants).

Green vegetables   The function of blood thinners (anticoagulants) can be altered if you drastically change your diet of green vegetables which are rich in vitamin K (i.e. spinach, broccoli asparagus etc.) and could cause blood clots. A change in your diet of liver can also affect the ability of blood thinners to do their job.

Fiber   Large amounts of fiber (i.e. oatmeal, bran, whole grains etc.) slows the absorption of many medications (i.e. heart medications, pain relievers or analgesics etc.).

Milk and dairy   If you have an allergy to milk or dairy (lactose) you should avoid bronchodilator capsules (Spiriva or Tiotropium bromide)

	due to the lactose in the capsule. Milk products can interfere with some antibiotics decreasing the effectiveness.
Over the Counter	Over the counter medications (i.e. herbs, antacids, iron pills, vitamins, cold/flu medications etc.) can effect the prescription medications you are taking. Ask your pharmacist for advice before you buy any, any over the counter medication.
Peanuts and soy	If you have an allergy to peanuts or soy you should avoid bronchodilator inhalers (such as Atrovent or ipratropium bromide and Combivent) which contains soya lecithin (a yellow-brownish fat similar to that found in egg yoke).
Tobacco	If you start smoking or quit smoking you medications may need to adjustment (especially true for inhaled medications and corticosteroids). Chemicals in cigarette and cigar smoke are removed by your liver and when you start or quit smoking the livers ability to function will change.
Tyramine	Tyramine, found in fermented foods, pickled or smoked meat products (except cured ham), wine, tap beer, yogurt, processed foods, bananas, pineapples, tofu, cheese and many other foods. Tyramine can increase your blood pressure and heart rate. Bronchodilators or antihistamines increase the heart rate and blood pressure and tyramine can increase this effect. This is especially true when taking medications for depression (i.e. MAOI's or Monoamine Oxidase Inhibitors).

Should you decide to change your diet (the food that you eat) you need talk to your healthcare provider before making any major changes. Changes in the type or amount of foods you eat will have an affect on your medications.

Note: Discuss these issues with your doctor and pharmacist. Take your medications as prescribed (i.e. at the right time, with or without food, etc.), never stop taking or alter the dose of medications prescribed unless directed by your healthcare provider.

Homeopathic Medications (Herbs and Stuff)

Homeopathic remedies have worked for thousands of years around the world and they still work today. Discuss them first and with your healthcare provider and work together to coordinate them with your prescription medications. Additionally, many states have licensed Naturopathic Doctors that you can consult. Vegetables such as onions and garlic can reduce bronchial constriction. Radishes and honey can relive a cough. Patients with asthma may benefit by eating cabbage however, they should avoid milk and dairy foods because the increase mucous production. Teas such as green tee can trigger an asthma attack. Always discuss your homeopathic remedies with your doctor and/or pharmacist due to medicine interactions.

You will find many sources for herbs or homeopathic medications on the internet and at your local health food store or public library. Always make sure your sources of medical information are trusted ones. Remember, the FDA (food and Drug Administration) does not regulate much of this information until there is a problem.

Note: Just because it is over the counter, it does not mean it is safe or without side effects. Talk to your healthcare provider or pharmacist before taking them. Always look for the U.S. Pharmacopeia (USP) label or stamp on those over the counter (OTC) medications. The U.S. Pharmacopeia is a voluntary testing program that certifies the contents on over the counter herbs and supplements.

Oxygen

Oxygen is a medication. If you find it is often necessary to increase or decrease your oxygen flow rate, tell your healthcare provider, this could be an indication of a changing health status. If you are at home fighting a small bout of the flu or a cold and need more oxygen, tell your healthcare provider. This is true with all mediations.

Contrary to popular belief, oxygen is not flammable (it does not burn or explode). Oxygen supports combustion; it is an oxidizer and makes anything flammable, burn really hot and fast, almost at an explosive level. Therefore, "Oxygen and an open flame and even extreme heat can be a dangerous combination".

Oxygen, being an oxidizer (supplies oxygen for rusting and burning) will react with petroleum. You may wonder why we mention this; oxygen is generally dry and can cause chapped lips and sensitivity to the nose. Oxygen will react with any product that contains petroleum and cause your chapping to become worse. You should use a moisturizer that is petroleum free. We have been fighting over the use of petroleum products with hospitals for years, but the patients that have taken our advice have commented about how much it has relieved their suffering.

Pharmacology

Drugs or medications are named by their brand name, generic name or chemical name. This can be very confusing, so do not be afraid to ask your healthcare professional or pharmacist for clarification. You are first in line in the prevention of medication errors.

Pharmacology is the study of drugs and their interactions with living organisms. Medications come from plants, animals, minerals or may be synthetic. They are used for three reasons; preventative, therapeutic (correct or cure) and diagnostic needs. Notice that recreational is not listed.

Using the pharmacy

Try to use only one pharmacy.

Some medications interact with others, greatly increasing their effect (potentiating or synergy) or greatly reducing their effect (antagonism). Sometimes these interactions can be life threatening. The pharmacist understands this and even has computer programs that will alert them to potential problems. The pharmacist can even alert you to foods that will affect the medications. If you receive some medications by

9

mail and some from your local pharmacy, make sure both suppliers are made aware of all of the medications that you are taking.

Pharmacies located outside the control of the FDA (i.e. Canada, Mexico etc.) may not have the same quality controls as pharmacies within the USA.

The dosages may not be correct or mixed with something not approved by the FDA or it may be the wrong medicine entirely. They are cheaper for a reason, so be careful. We are acutely aware and tremendously sympathetic to the nearly insurmountable financial burden that many Americans face in securing their much needed prescription drugs. There is no doubt that this situation deserves immediate attention but the possibly of jeopardizing one's health by securing non-FDA approved drugs, is something that must be carefully considered ("Shades of Catch 22").

Note: There are many programs to assist you in obtaining your medications, ask your healthcare provider for assistance and guidance.

Always take the medications as prescribed. Reducing or increasing your dose and/or frequency could kill you. It could cause side effects or even make your condition worse. Complete your antibiotic regiments as prescribed. By stopping the medication when you feel better, you may be allowing the bacterium that survives to develop a resistance to the medication or allow it to re-infect you soon after stopping the medication. You must taper steroid dosages because the side effects can be significant. You must believe that your healthcare provider has your best interest at heart.

When you age given a new prescription medication, one never taken before, ask the pharmacist why the medication was prescribed. This question will help eliminate errors with prescription medications (we will cover this in detail in under medications). Additionally, ask about the medication's side effects and how long each dose lasts (this will allow you to plan your life better).

Generic Prescription Medications

Everyone knows that generic drugs are cheaper than name brand medications (up to 80% less). However, generic drugs are not identical to the name brand counterparts. The FDA (Food and Drug Administration) requires only that the active ingredients be the same. The medication (especially extended release) can release or breakdown in the body differently resulting in the medication not performing properly. Talk to your healthcare provider about your specific medications before you make the change to generic medications. Saving money is a great idea as long as you can live to spend your savings.

If you change to a generic medication and you do not feel right, you may develop new side effects, or the medication may not work as well. You need to discuss this problem with your healthcare provider; if you do not tell them, they will never know there is a problem.

Note: Over the counter (OTC) medications can create the same issues since they are not exactly the same. Ask the pharmacist if you have any concerns.

Hospitals

If you have ever walked into a hospital room, I am sure that you noticed a myriad of monitors with lights, numbers, and other heart monitoring features flashing above the bed. Do not be afraid to ask what each of these are, there are no secrets. They are all there to monitor your loved one's vital signs and to notify staff to any problems that arise. Understanding the things that you seen and hear will make your hospital stay much more comfortable.

Note: If the hospital or facility does not look clean, you need to find a cleaner place to stay. This is a good indication that they are not a high quality facility.

Patient Rights

You have the right to be treated with curtsey, respect, dignity, and privacy.

You have the right to receive emergency treatment or medical care, regardless of race, ethnic group, religion, handicap or your ability to pay for the care. Additionally, you have the right to know how much treatment will cost you before the care is given (if requested).

You have the right to be informed about your illness or treatments no matter what your primary language and in the terms, you understand.

You have the right to make informed decisions concerning your healthcare and the right to appoint someone else as your surrogate to make decisions for you (see chapter 4).

You have the right to choose your home medical care company, pharmacy and other healthcare providers.

Every hospital and faculty has rules and if you are interested, they must be supplied to you.

You can inform the facility, at any time, if you are not satisfied with your care. Do not be afraid to help the facility solve their problems. You will always be welcomed back when you need them.

Patient Responsibilities

You are responsible to give the facility correct and complete information.
You should maintain written copies of medication and any legal papers (see chapter 4) the hospital or facility may require (i.e. living will, advance directive, surrogate information, powers of attorney etc.).

You must confide in your caregivers if feeling different or worse.

You must follow the direction of your healthcare providers.

Visitors must follow all facility rules.

Your bill must be paid promptly or you must contact the facility to make payment arrangements.

Hospital Discharge Plans

Hospital discharge planning is just as important as your stay in the hospital. Your healthcare provider, nurse and respiratory therapist need to make a collaborative effort, with their focus on you and your care. The day you are admitted to the hospital, you need to tell your doctor you will require a discharge plan to be given to you and discussed with you prior to leaving the hospital. You may need nursing, respiratory therapy, physical therapy and/or other types of assistance in your home. Once you are home, it will take time and great effort to get answers to your questions. Set yourself up for success.

Remember you are the customer, and as an informed medical consumer (medical advocate), you are your own first and last line of defense. Many elderly people are afraid to ask questions, or to question something that makes them uncomfortable. The doctor and hospital will always welcome you back, so stand up for yourself and for your love ones.

Breathing Retraining

For patients with COPD (Chronic Obstructive Pulmonary Disease) or other lung disease, dyspnea (shortness of breath or difficult breathing) instills a fear of suffocation. The resulting anxiety can be a viscous circle of progressively worsening symptoms. The fear or anxiety develops into panic, which increases the heart rate, increases dyspnea, increases blood pressure and increased your rate of breathing. This increased the demand for oxygen making the dyspnea worse and the panic deepens making you worse and worse until it is a real medical emergency. The cycle can only be broken by medicating (sedating) the patient or by the patient controlling the anxiety and not allowing it to become a panic situation.

Relaxation Training is a tool you can practice prior to a panic attack in order to help minimize the physical effects of your experience. This is training you can do almost anywhere. Practice tightening muscle groups like shoulders and arms, legs and feet. Lift your shoulders and relax them or lift your toes and relax them and repeat. These are simple and sound dumb, but they work.

Biofeedback is another relaxation technique you can practice prior to any anxiety attack. It is all about helping you relax. By finding your happy place, thinking happy thoughts or concentrating on something that makes you feel secure you will relax. You can verify this technique is working by checking your heart and breathing rate. If you own a pulse oximetry meter (the red finger probe) you can use that instead of a watch or clock. Again, we know it sounds dumb but it will work for you.

Breathing Exercises

Diaphragmatic breathing exercises help you to strengthen your diaphragm (the muscle that moves air into and out of your lungs), expand your lungs so that they take in more air, and reduce the work of breathing (WOB). When you experience difficult breathing or shortness of breath (dyspnea), in addition to using your diaphragm, you will use accessory muscles in your chest wall, back and neck muscles in order to breathe. All this work will exhaust you and it will become a medical emergency. Through strengthening your diaphragm muscles, you will decrease the work of breathing and better control your breathing. Here is how to do the exercise:

First lie on your back (supine) with your knees bent (about 90 degrees) with one hand on your stomach and the other on your chest.

Inhale slowly through your nose. Movement should over your stomach only, the hand over your chest should not move at all.

Tighten your stomach muscles and exhale using pursed-lip breathing. You should feel no movement over your chest while exhaling.

Once you have mastered the procedure you can learn to practice the procedure while setting.

You should lie on your back (supine) as possible, with one hand on the abdomen and the other hand on the chest. While taking a breath in, press on the abdomen and extend the abdomen outward as far as possible during inhalations. The hand over the abdomen should move outward while the hand over the chest should not move. Once the patient has mastered the procedure while supine, they should practice doing it while sitting or standing. A good exercise program should last 20 – 30 minutes 2 or 3 time each day (BID or TID).

Pursed-lip Breathing Exercise is a simple exercise you can practice anytime and anywhere. This prolonged exhaled breath with the slight back pressure from your lips opens your small airways and allows old air to come out and fresh air to reenter your lungs. The exercise will allow you to relax, will decrease shortness of breath (dyspnea), decrease your effort in breathing and may extend your activity level. This is how to perform the procedure correctly:

First, relax your shoulder and neck muscles.

Inhale through your nose normally and slowly (about 3-5 seconds).

Press your lips together (as if you are going to whistle) and exhale gently (for about 4-6 seconds).

By practicing and utilizing breathing exercises and relaxation techniques you can greatly reduce the frequency and severity of dyspnea associated with suffocation anxiety or panic attack. Through practice, you can make their use "second nature" (you will not even notice you are doing them) and this will make a tremendous difference in a respiratory emergency. These are things you can do on your own to help yourself or practice with a loved one.

Exercise and Fitness

Fitness and exercises (about 30 minutes total per day) will help you heal, get well and improve your quality of life. Even a little exercise will go a long way in helping you. A tailored program of stretching, aerobic exercise and resistance training designed to fit your needs and physical restrictions is best. Remember, prior to beginning any exercise or fitness program consult your healthcare provider.

Shoulder Shrugs strengthen shoulder and upper back muscles.

Sit with feet slightly apart, arms at your sides and shrug (lift) shoulders up and relax.

Arm or Elbow Circles will strengthen arm and shoulder muscles.

> Stand or sit with hands on top of shoulders or out stretched.

> Move arms in a circular motion.

Shoulder Circles

> Sit with hands at your sides and make circular motions with your shoulders.

Head Circles will strengthen your neck muscles.

> Sit and roll you head from side to front to side slowly. Try not to roll your heal backwards.

Overhead Shoulder Stretches will strengthen your arms, chest and upper/lower back muscles.

> Sit in a good stiff chair with your legs spread slightly.

> With your arms outstretched and relaxed slowly raise your arms as far as is comfortable.

> Slowly lower your arms and repeat.

Hamstring Stretches will loosen and tone your lower back and hamstring (back of the upper leg) muscles. This is like touching your toes however; you will be using two sturdy chairs.

> Sit in one chair and place you foot on the other chair

> Slowly bend forward while keeping your back strait is possible. Do not bounce or force yourself forward to far.

> Hold this position for a few seconds and repeat.

Quadriceps Stretches will stretch and strengthen your quadriceps (the big muscle on the front of your upper leg).

> Stand or lie down. If you stand, use a sturdy chair or a secure railing for support.

> Hold your ankle or pant leg and pull your heal to your buttock and hold for a few seconds.

> Repeat with both legs.

Calf Stretches will stretch and strengthen you calf muscle (big muscle on the back of your lower leg).

> You will need something to hold on to for balance and something that is elevated 3 to 10 inches (like a stair step).

Place the front half of your foot on the step and lower your heal and hold for a few seconds.

Repeat for both legs.

Note: Stretching, resistance training and aerobic exercise can be fun as well as beneficial to your health, however always ask you healthcare provider prior to beginning any new program.

Forceful Coughing will help you cough up the mucus trapped in your lungs and trachea (breathing tube). It is best to use a group of two coughs. The fist will move the mucus out of your lungs and the second to get rid of the junk. This is how to perform the exercise:

While seated comfortably with your feet on the floor, lean your head slightly forward (do not over-extend your neck).

Inhale deeply using your diaphragm and hold your breath for at lease three seconds. Cough as hard as you can to start the mucus moving, and then cough again. Repeat as necessary taking breaks between attempts.

Nutrition

Nutrition is as important as your medications, equipment and exercise. Reducing fat in your diet will reduce bad cholesterol (LDL). Talk to your healthcare provider about foods you should add to your diet and foods that can affect you medications or health. Many fruits, berries and juices (contain antioxidants, vitamins and minerals) can make a tasty smoothie. Remember what your parents taught you; the proper diet (not dieting) will prevent you from ever being required to diet. It will make you healthier and help to keep you healthy for a lifetime. The food pyramid is the key to a proper diet (you may want to buy a copy of a book titled The Prescription for Wellness); it may help you eat well.

Your body requires some intake of fats to allow it to use various vitamins and minerals. Talk to your healthcare provider about how much fat your body needs before cutting them out of your diet altogether.

Always discuss diet changes with your healthcare provider prior to making any major changes in the foods you eat or drink.

Stress Management

Stress can cause you veins and arteries to tighten up (or spasm), increase your blood pressure and increase your heart rate. There are hundreds of simple stress management techniques (i.e. meditation, aroma therapy, yoga or just setting and relaxing, to name a few) that would greatly benefit to your health when you reduce stress (about an hour per day).

Note: When you combine exercise, healthy nutrition and stress management you can stop the progression of diseases and repair some damaged (like heart disease and clogged arteries).

Sleep Hygiene

As we age (50+), millions of us will develop sleep problems. Sleep is as important as diet in promoting good health. Lack of sleep can have many causes; sleep apnea (stopping breathing for >10 seconds), stress (work or health), Restless Leg Syndrome (RLS), traveling (time zones), worry (money, family etc.), and/or insomnia are some of the most common causes for lack of sleep. Insomnia has three different types; Transient Insomnia (over short periods), Intermittent (on and off), and chronic (most nights, lasting for a month or more).

Lack of sleep causes real problems; cardiovascular disease, headaches, memory loss, tiredness, lack of energy, inability to concentrate irritability, depression, anxiety and weight gain. Additionally, a real danger of falling asleep while driving or operating dangerous equipment during the day exists. As you can see, it is of upmost importance to diagnose why you do not sleep well and develop a plan to resolve the problem.

Adopting healthy habits, a healthy lifestyle and making your bedroom "sleep friendly" are crucial to good sleep hygiene. Below you will find tips on getting a better nights sleep:

Avoid caffeinated drinks and foods (cola, tee, chocolate, coffee etc.) within six hours of bedtime.

Do not nap longer than 30 minutes during the day to facilitate sleeping at night.

Schedule routine sleep and wakeup times (including weekends).

Discuss with your healthcare provider any medication you take that can cause insomnia. You may decide to take these medications early in the day.

Exercise routinely (morning or afternoon) at least 3-4 hours before bedtime.
At night you should not lie awake in bed for more than 20-30 minutes. It may be better to get up and try again after you have a chance to relax. Often events of the day can make it hard to relax enough to fall asleep.

Alcohol can help you get to sleep, but most people wake up during the night and have trouble getting back to sleep.

Do not sleep with the television or radio on (except with soft music and no commercials).

Develop a comfortable environment for sleeping. Remove nosy or bright clocks, night lights (light signals your brain that it is time to wake up) , wear comfortable clothing and use comfortable bedding.

Try not to eat spicy foods too close to bedtime because they can cause discomfort. Reduce the amount of liquids you drink at night to eliminate the need to go to the bathroom in the middle of the night. A light snack can prevent hunger during the night allowing you to sleep better.

Develop a relaxing routine such as reading, soft music (CD or tape), relaxation techniques, warm bath, comfortable room temperature and/or gentle massage.

Sleep disorders significantly affect your quality of life, health and safety. They can also lead to severe headaches. Questions to think about that may indicate sleep problems are as follows:

Are you tired during the day?

Do you ever fall asleep during meetings, at family events, driving or while watching television?

Have you been told that you stop breathing, snore or move around a lot while sleeping?

Do you drink many caffeinated beverages (i.e. coffee, tea, soda or energy drinks) during the day?

Are you taking medications late in the day that conflict with your sleep (some medications can keep you awake)?

Are you and your partner happy with you quality of sleep?

Keeping a sleep record or diary (7-30 days) will help your healthcare provider get a better picture of your sleep patterns or problems. You should include your habits, sleeping pattern, diet, daily spousal comments and things that bother you during the day, which might affect your sleep.

Medications prescribed by your healthcare provider may help however; you are at risk for falling and confusion if you need to get up at night for whatever reason.

Polysomnography (sleep study, see chapter 11) may be instrumental in determining causes for your sleep interruption. It may also allow you to get a better night's sleep. This test could save your life or marriage (depending on how loudly you snore).

The National Sleep Foundation website (http://www.sleepfoundation.org) has additional information. This organization is dedicated to improving sleep patterns, decreasing sleep problems, and identifying health and safety issues. Your healthcare provider will appreciate your active involvement and you will feel much better about any testing procedures they prescribe when you understand the benefits.

Rehabilitation

Rehabilitation is a bad word. Who ever decided Rehab describes what is accomplished was dead wrong. Rehabilitation often makes patients feel they have done something wrong. Maybe it should be renamed reconditioning or happy fun time so that does not sound so bad. In fact, everything you have read so far in this chapter is part of your rehabilitation program. Many healthcare systems are developing Advanced Illness or Coordinated Care Programs to support patients facing an advancing illness. Ask your healthcare provider what programs are available to you and your family.

Note: Rehabilitation will get you back on your feet. It one of those things you just have to do.

Pulmonary Rehabilitation

Pulmonary rehabilitation works best as a structured program utilizing goals that a patient sets, achieves, and continues on their own. Your healthcare provider as an individualized program consisting of prescribed exercise, education/training, psychosocial assessment and outcome assessment lasting up to 72 sessions initially prescribes it. Additionally, heart rate/blood pressure, arterial blood gasses (ABG's), bronchodilators with spirometry (before and after), flow volume loops during exercise studies with pulse oximeters are helpful (see chapter 11). These tailored programs begin with the collection of data about condition and goals in order for a detailed rehabilitation plan/program to be established. Many patient goals may be similar such as to decrease dyspnea (difficulty breathing), to develop the skills necessary to properly administer medications, disease education, increased strength and endurance, stress management, improved lung function and the ability to recognizing early symptoms etc). Everyone is different; you may require a more personalized assistance or guidance than does someone else. It is important to remember, the only dumb question is the one that is not asked, so "don't hesitate to ask for the help you need". Patients must commit to finishing a program, and if ongoing, to continuing the program (maybe for life). Your motivation and dedication to the improvement of your quality of life are the most important factors in becoming stronger, physically and emotionally. Optimism may be the major component in your improving your quality of life. This is where a "personal pulse oximetry" will come in handy. Do not focus on low oxygen levels but rather on increasing the effectiveness of each of your breaths to improve your oxygen level.

A pro-active (or what can I do) and responsible attitude (it is my problem and I must do what it takes) must be taken concerning your health.

Cardiac Rehabilitation

Participation in a cardiac rehabilitation program (which has three or more phases, hospital, home and outpatient) can significantly lower your risk for future heart problems. It will improve your health and increase your quality of life. As with Pulmonary Rehabilitation, you help set the goals specifically designed for you and your medical needs. This in turn will help you to be able to do more of the things that you want to do in life.

Your rehabilitation program begins in the hospital with lung exercises (such as the incentive spirometer or IS), airway clearance or coughing techniques (splinting your chest with a pillow or stuffed animal), and breathing techniques (pursed lip breathing).

How fast you recover will be based your attitude (how seriously you take your participation and responsibility in rehabilitation), age and health (other conditions or disease that will slow your recovery),

Your healthcare team will assist you in the following areas and more:

> Education Support for the thing you and your family need to know. Things like lifestyle changes in nutrition, recommendations for developing healthy habits, understanding proper medication use, proper exercises, your safety, any limitations temporary or long lasting limitations you may experience and smoking cessation.

Emotional Support for all the things you have and are going to go through. They will instruct you in relaxation techniques and provide smoking cessation support.

Team Support in any groups and family areas as well as counseling for depression or anxiety.

If you should experience unusual discomfort (chest pain or angina), difficulty breathing (dyspnea), sweating, heart palpitations, nausea, unusually high or low blood pressure or swings in your blood sugar you should seek immediate medical attention.

Note: The combination of Cardiac & Pulmonary is Cardiopulmonary Rehabilitation.

Electronic Medical Records (EMR)

Electronic medical records (EMR) will allow (only if you authorize the release of the information) your healthcare provider to exchange your healthcare information with facilities everywhere. This means that if you have a medical emergency while you are on vacation in New York, Florida or California your medical records are available where you need them to support your care and your healthcare team at home has access to your care you received.

An electronic medical record also allows healthcare professionals to view your treatment history without thumbing through incomplete paper records or requesting medical record copies and automatically schedule preventative care throughout the year as required.

Electronic medical records will automatically interface with hospital departments (i.e. respiratory, laboratory, X-ray, etc.) and devices (i.e. EKG, breathing rate, blood oxygen levels etc.) which will reduce the chances of medical records errors and reduce your overall medical costs. They will also keep track of ionizing radiation exposure (i.e. X-ray, CT etc.).

Organ Donation or the Gift of Life

The decision to donate will never interfere with your medical care. Your decision is an important one and if required your family is consulted prior to your donation. The collection team is separate from your healthcare provider and caregivers. Under the law, a family cannot override your wishes, as long as you have completed the required documentation.

The recovery process, is a surgical procedure and will not change the appearance of the body. Donations are assigned to recipients based of body size, blood type, medical urgency and time spent on the waiting list. Wealth, gender, religion, age and race are not determining factors in who is to receive a donation.

The donation of heat, lungs, liver, kidneys, pancreas and intestine can save 8 to 10 lives. Tissues such as eyes, skin, bone, heart valves, veins and tendons can save over 50 lives and benefit almost 100 people.

The National Transplant Waiting List has over a dozen people added each day. To date there are over 100,000 people o the list. Sadly, each day about 20 people die while waiting for the transplant. Most religions support organ donation and it is a generous gift, it is a gift of life for so many. You will find legal documents and forms in chapter 4 that will help guide you with respect to the paper work.

Deep Vein Thrombosis or DVT

DVT is a blood clot that normally formed in your legs. This type of clot is caused by inactivity (i.e. setting on long plane/train/automobile/bus rides, wearing casts, etc.). If a clot is dislodged (breaks free), you have just spun the roulette wheel of death and destruction. If the clot is stuck in the wrong place your brain or heart could be seriously damaged (see chapter 9).

Smoking

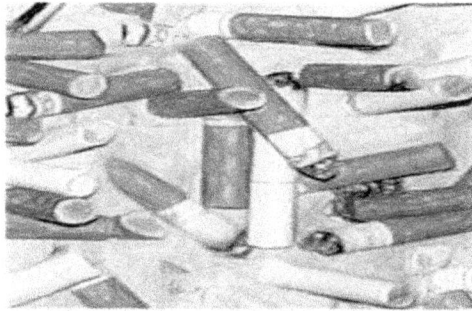

Cigarettes

People breathe 8 to 20 times per minute (normally) which is about 9,000 to 17,000 liters per day (could be more). Normally we inhale about 530 million particles of offensive stuff each day (dust, allergens, microbes etc) from the atmosphere and smokers may more than double this amount. Mucus is one of the defenses that our bodies use to remove these particles and smoking tends to overload the whole process.

Smokers need to understand that cigarette products such as "light" or "ultra light" are not really reducing the amount of tar and nicotine inhaled. These cigarettes have ventilation holes to add air as you inhale. Blocking the ventilation holes or inhaling more deeply, puffing faster or smoking more cigarettes yields more tar and nicotine. There are no safe cigarettes, unless you buy them and quickly place them in the trash.

Public health departments have concluded that secondhand smoke from cigarettes causes' disease. If you smoke in your home, you could be causing loved one to be sick as well, including lung cancer, heart disease, asthma, respiratory infections, coughs, wheezes, ear infections and/or Sudden Infant Death Syndrome. It also makes it much more difficult to sell your home. If you have to smoke, do it outside, it might be the first step in quitting.

The FDA (Food and Drug Administration) is now in charge of regulating the ingredients in cigarettes. Maybe someday it will be easier to quit however the sooner you quit the better off you will be.

Smoking Cessation

What else can be said about the health risks and the costs of smoking? It is on every pack of cigarettes, it is on TV/radio, and it is everywhere. Over 400,000 Americans die every year from tobacco related diseases and it decreases your life expectancy by 10-12 years. Healthcare costs and loss of job productivity costs us $100 billion each year. Everyone hates being told what to do by the government however; we all know smoking is bad for us and maybe even worse for others (second hand smoke). Even with states passing laws restricting public smoking, more and more people start smoking every year.

Well you know all that. The hard part is quitting and more so, to stay quit. All we can do is give you the information and hope you are strong enough to be successful. Here are some ideas to consider:

Talk to your healthcare provider about medications to help and call around town for a Smoking Cessation Class (most hospitals have programs). Look for help on the web, most states have programs to help. Many state programs give away gum, patches, lozenges or sprays to help you quit. Call the American Cancer Society and the America Lung Association. Try over the counter nicotine supplements (i.e. sprays, Patches, gum, lozenges, herbs, etc.). Start your program with friends, loved ones or a group for support.

Develop a personal support group in order to boost your chances of success.

Ready: Select a date to quit and mark the calendar. Write down the benefits of quitting and keep them in your wallet or purse. Throw away ashtrays and other reminders of smoking. Review previous attempts of quitting, what things made it difficult for you. Now that you have identified what people, places or things that caused you to fail you must avoid these situations for a while.

Support: A support person or group will increase your odds of success. Discuss why you want to quit with these people. Your healthcare provider is a great source of help.

Behaviors: Avoid temptations (people, places or things), find new activates that do not include smoking. When the urge to smoke strikes, consider exercise or anything that will help you quit. Discuss this with your healthcare provider first and do not over do it. Keep your hands and mouth busy. Nicotine gum/lozenges, healthy snacks (i.e. carrots, celery, water, etc.) can be a great help. Call a support person for extra strength.

Medications: Your healthcare provider can assist you with medications that will help you quit smoking (Zyban®, Wellbutrin® or Chantix® to name a few). Do not barrow medications from a friend because only your healthcare provider knows if one of these medications is a safe option for you. Other over the counter nicotine replacement products are gum, patch, inhaler, nasal spray and lozenge may be helpful to you. Anti-nicotine vaccines in clinical trials may be available soon. The vaccine creates antibodies that are attracted to nicotine like a magnet (or have an affinity to nicotine) making the nicotine too large to penetrate the brain blood barrier (too big to get to your brain). This will be another option to discuss with your healthcare provider.

Note: Electronic cigarettes or e-cigarettes are unproven as nicotine replacement therapy and as of late not been evaluated by the FDA. There are concerns concerning the possibility of increased nicotine addiction, untested ingredients, as well as the overall safety of the chemicals they contain. Talk to your healthcare provider first.

Benefits in smoking cessation start immediately, within 30 minutes your heart rate will slow. Within the first 24 hours, your carbon monoxide (CO) level will decrease to normal. Within 1 to 10 months your shortness of breath (dyspnea) will decrease, your cough will subside and your risk of heart attack (MI) will decrease. During the next 5 to 15 years, risks for stroke and cancer will reduce by half. Quitting will slow the progression of COPD (Chronic Obstructive Pulmonary Disease). Remember it is never too late to quit.

It is not easy, in fact, it is very hard to quit, but here are many benefits and maybe some of the following will help you keep focused on your goal:

Reduce risk of cardiovascular disease by 30%
Increased self-esteem.
Your voice will improve.
Reduced risk of emphysema and cancer.
You will sleep better and smell better.
Reduce the fire hazard.
You will smell better.
Reduce the risk of ulcers.
Your sense of smell and taste improve.
Reduce periods of shortness of breath.

Improved sense of self-control.
Asthma attacks will be less frequent.
Reduced risk of stroke and heart attack.
You will live longer and save money.
You will have better circulation.
They cause wrinkles so you will age faster.
Improved strength and endurance.
Experience fewer colds during the year.
You have more time to do stuff.

It may help to write down the benefits of quitting vs. what you like about smoking (yes what you like) and keep the list with you to review when you begin to lose control.

Find a small jar, one with a lid (like a baby food or jam jar) and use it as an ashtray for your last cigarette or two. Place it out of reach of children on a shelf and save it for a while. When you feel you are loosing the battle not to smoke, remove the lid and smell the contents deeply. That should help, it really does stink. Put it back on the shelf, you may need it again.

Remaining smoke free is the hard part. According to the Centers for Disease Control and Prevention, you are at greatest risk for relapse during the first three months after quitting. If you do break down and smoke, do not give up. Remember what caused your program breakdown and learn from your momentary lapse. Compare the reason for your lapse with your "want to quit list" in your purse or wallet. Adjust what you do daily to avoid the situation responsible for the lapse.

You may want to sit down and figure out how much money you are saving. If you were smoking one pack per day costing $9.00, you will save almost $3,300.00 each year on cigarettes alone.

Good luck, we know from experience it is not easy, but staying quit does get easier over time.

Problems with PAP (Positive Airway Pressure) Equipment (see Chapter 6)

Many common problems with equipment at home have very simple solutions (called troubleshooting). Always remember to be safe; check power cords for problems and do not forget to check the breakers if your system will not start.

Often basic understandings how things work help you correct problems and keep thing running smoothly (see chapter 6).

Note: Never open electrical equipment to find out what is wrong.

Problems and Solutions		
Problem	**Cause**	**Solution**
Air Leaks	Old equipment is worn or stretched	Replace parts
	Improperly adjusted head gear	Adjust headgear
	Improper fit	Change mask
	Hair interfering with seal	Reposition, change to nasal prongs, or shave
	Pressure set to high	Check PAP setting or consider BiPAP
Chest Discomfort	High airway pressure or swallowing air	Decrease PAP setting or consider changing to Auto PAP or BiPAP
Claustrophobia	Anxiety	Desensitization training (see chapter 6) or change from the mask to nasal prongs
Dry Nose, Throat or mouth	Dry air on mouth leak	Add a heat/humidifier, chin strap or consider a full face mask
Ear Pain	High airway pressure or nasal congestion	Verify PAP pressure, consider Auto or BiPAP, consider using decongestants or steroids
Eye Irritation	Air leaks	Readjust head gear, consider a different style of mask/nasal prong, also see Air Leaks
Nasal Irritation or Congestion	Dry air or nasal allergies	Consider Heat/Humidification, nasal steroids, decongestants or contact your healthcare provider
Gastric bloating	High airway pressure or swallowing air	Decrease PAP setting or consider changing to Auto PAP or BiPAP
Skin Creases	Head gear too tight	Readjust head gear, consider changing the style or size of mask/nasal prongs
Skin Irritation	Sensitivity to materials, improper adjustment, heat rash	Consider changing to nasal prongs, Readjust head gear, decrease heat/humidifier temperature, use a skin protection, or clean equipment

For equipment information, see chapter 6

Learn CPR (Cardiopulmonary Resuscitation)

The most important thing that you may ever do is to learn CPR. This CPR should be a required class every year. It should be a prerequisite to graduate high school or college and when being hired for a job. We have personally saved five neighbors with CPR and anything that you can do to help a heart attack, choking or drowning victim is better than doing nothing at all. The Good Samaritan Law protects you when you try to help someone even if you do some things wrong. If you a worried about germs, with the new procedures for CPR you do not even have to put your mouth on theirs anymore. We have included the new procedure basics in the chart below:

Hands-Only CPR

The American Heart Association has announced that 100 compressions per minute uninterrupted until paramedics or medical professionals are able to take over even without giving mouth to mouth breaths can save a life. So if you are concerned about placing your lips on someone during CPR, don't. Your giving compressions only could save a loved one, friend or stranger.

Untrained Person Performing Hands Only CPR

Maneuver	Adult	Child	Infant
Rescuer	Adolescent - Adult Age Over 8 to Adult	1 year - Adolescent Age 1 year to 8 years old	Under 1 year
Activate EMS Call 911	If suffocation arrest compressions for about 2 minutes then call 911	5 cycles of 30 compression followed by 2 breaths then call 911	5 cycles of 30 compression followed by 2 breaths then call 911
Circulation			
Compression Landmarks	Center of chest just below the nipple line		Between the nipple line
Compression Method	2 hands, heel of 1 hand over the other (Do not bounce)	1 hand, heel of 1 hand	2 fingertips
Compression Depth	2 inches	A third to half the depth of the chest	
Compression Rate	Approximately 100/min Continue compressions until help arrives	30 Compressions then 2 100 breaths approximately compressions per minute until help arrives	

Note: During a cardiac emergency, your chest compressions give the medical professionals the time they need to deliver the electrical shock (defibrillation) it takes to save a lfie.

Trained Rescuer CPR

Once you have taken a CPR class you will be better prepared to save the lives of loved ones or even a not so perfect stranger. CPR Training enables you to recognize a choking victim, a heart attack victim or a person in respiratory arrest (stopped breathing). In the case of a heart attack victim, your actions prolong the time the victim has to receive defibrillator (electric shock that resets the heart's electrical system). Any person who can obtain an Automated External Defibrillator (AED) can use a defibrillator. What most people do not realize is that the electrical shock from the defibrillator is what actually saves the life of the heart attack victim.

Heimlich Maneuver

The Heimlich Maneuver is also known as abdominal thrusts. The goal is the same as the procedure contained in CPR above, to remove foreign objects from the trachea (windpipe). Using this procedure the patient can be standing or setting by hugging the patient from behind, making a fist with both hands clasped together and making an inward and upward forceful compression just under the ribs. The goal is to create pressure under the obstruction, forcing it upwards and out of the mouth like an air cannon.

We know this review seems quite complicated but once you have taken a CPR class, you will view it much differently, more like a reference card or a reminder of the things that you already know. Investing a few hours in order to save someone's life, maybe even a loved one is really worth the effort. Please take the time to learn CPR.

Note: Everyone should have a basic emergency medical base of knowledge. Take the time to take a First Aid Course, First Responder Course and/or a CPR course.

Other Information Resources

Web based medical information site http://www.webmd.com.
Aging with Dignity http://www.agingwithdignity.org.
American Association of Retired Persons (AARP) http://www.aarp.org.
Local hospital, nursing home, hospice, or home healthcare provider.
FAA & DOT regulations http://www.regulations.gov.
National Sleep Foundation website http://www.sleepfoundation.org.
National Institute of Health http://www.nih.gov.
Wikipedia at http://en.wikipedia.org.
Drug Information Online http://www.drugs.com.
World Health Organization http://who.gov.
Air Quality Index http://www.epa.gov or http://weather.gov or http://www.airnow.gov.
American Association of Respiratory Care http://www.aarc.org.
The American Herbal Products Association, http://www.ahpa.
The national Library of Medicine, http://www.nlm.nih.gov.
The United States Pharmacopeia, http://www.usp.org.

Your local hospital, healthcare provider and pharmacy usually have information available to you.

Please make sure any information source on the internet is a trusted site. You can find a lot of misinformation out there.

Be smart and be safe.

Notes: We have not seen you make any notes yet, so we made on for you.

Your flu shot will not make you sick, ask the doctor why.

Notes:

26

Chapter 2
Questions for My Doctor

When you are ordering office equipment or purchasing a new car, you do not ask questions of a general nature. You need the facts in order to be able to make an informed decision. Is your health, the health of a parent, or any other loved one any less important? Your answer should be emphatically "NO"!

Communications with healthcare professionals

We covered a lot of this in chapter 1 but it is important and we will build on what you now know. If you show that you are interested in your medical care and eagerly want to be an active participant of your healthcare team, it will go a long way in insuring you get the attention you deserve.

There are many questions you may need answered by your healthcare provider. The 5 most important questions are:

What do you think this could be?

What do we hope to learn from this test and will it change how you will treat me?

What outcome do we expect by taking this medication?

Do we have any other options?

What do we do next?

Note: When you are sick, you will not remember to ask the right questions, so write your questions down and bring this book along with you.

All healthcare professionals are busy and strapped for time, but all healthcare professionals are available and approachable to assist you. It is their duty. Your responsibility is to slow them down and get your questions answered in words you understand. If anyone on your healthcare team uses abbreviations or terminology you do not understand stop them and make sure you understand (take notes). Remember it is your responsibility to make sure you understand and do not forget to take notes (you will not remember when you get home). Do not be afraid to ask questions because of their title (i.e. Doctor, Nurse, Respiratory Therapist, etc.). You have the right to understand your healthcare plan.

Again, do not be afraid to ask direct and pointed questions. In our experience, it would be a welcomed and pleasant surprise for a patient to demonstrate an intense interest in his or her own wellness. Most people either do not know what to ask or are just afraid of the answer. It is important for you to know what Mom, Dad or a loved one likes to do on a daily basis or in their spare time. Armed with this knowledge you will be able to ask questions similar to the following:

Will mom/dad be able to walk to the mailbox each day?

Or

Will mom/dad still be able to attend church on a weekly basis?

Or

Will mom/dad be able to do the thing/things they love to do each day?

It is equally important to understand their wishes concerning life-support equipment. Mechanical ventilators and other life-support equipment designed to extend life will do just that but will in most cases, greatly limit the quality of life that the patient is able to enjoy. Try not to use the words could or maybe. Ask the doctor or nurse questions like the following:

Will mom/dad require long-term mechanical support?

Or

How is this support going to affect the rest of mom's/dad's life?

Or

What would you do if it were your parent?

Everyone at some point must make "tough decisions" concerning a loved one's healthcare. The optimum scenario would be to have, prior to a catastrophic event, consulted with that loved one, and to have become aware of their quality of life desires, demands and/or spiritual beliefs.

You need to ask pointed questions about the disease or process affecting your health. Avoid words like if or maybe and do not be afraid to ask the scary questions like:

What will be my restrictions or limitations?

Or

What can I do to help myself?

Or

What can family or friends do to help me?

Or

What is the outlook or how long do I have?

Most people have legal forms or documentation concerning their desires (see chapter 4). Preparation for asking the right questions will make things a whole lot easier. We view artificially extending life like this, if the doctor is going be able to fix the problem, we can handle treatment modality needed. We just do not want to spend the remainder of our life developing bedsores and wasting away on life support equipment. We have taken care of many patients in "vent farms" who never get any visitors and look very unhappy. After all the medical care is given and all that is left is death, this is not the way we would choose to spend our days. If you are having difficulty making quality of life decisions, visit a facility of this type, or one comprised of patients with very little hope of recovery.

Remember, knowing the desires of mom and dad is what is important. The type of medical care they desire and what they like to do with their "free time" and exactly what type of existence they would consider intolerable are their desires not yours. Ask them today. Do not wait until it is too late. If you do, you are going to be forced to make some hard decisions on your own. It is hard, but please read and discuss chapter 4 with them today.

Ask your healthcare provider if the pneumonia shot or immunization is right for you this year.

Make notes in this book or keep a Medical Diary, list your questions and take it with you to the doctors. He or she will be impressed with you interest and determination for understanding your health.

Medical breakthroughs happen every day. Researchers are working on new vaccines, testing procedures and treatment protocols right now. One of the new blood tests detects cancer in its initial stages (this test is awaiting approval and is quite expensive) but they are working on the problem. It is your responsibility to research and ask your healthcare provider pointed questions.

Notes:

Chapter 3
Hazards & Dangers In My Home

Emergency Medications or Rescue Medications

Emergency medications should be marked in such a way that anyone assisting you will know immediately what medications you require. Place them in a bag or container marked in red "Rescue Drugs" or "Emergency Medications". Additionally, an EMERGEMCY LIST (see chapter 4) of all medications should be hanging on your refrigerator for EMS personnel.

When admitted into the hospital, a special procedures clinic or any medical facility, always check your armband to ensure all the information is correct. The wrong name, blood type or identification number could mean you may not get the right medications, blood or procedure. Healthcare professionals are people too, they can make mistakes, and you are your last line of defense.

Medication Safety

The medications and the doses ordered by your doctor are important. Use the medications the way it was prescribe for you. If you have changed your dosing from what was prescribed by your doctor tell him/her and if necessary, they will change the order so you do not run out of medications or hurt yourself.

Medications have a shelf life. Do not use them after they have expired. It is just not a good idea. Some medications such as antibiotics can become poisonous over time so, do not save them take all or throw them away.

Medications names include their brand name, generic name or chemical name. This can be very confusing and could result in overdosing. Many prescriptions and over the counter medications contain more than one medication. It is your responsibility to read the ingredients to prevent an overdose. Do not be afraid to ask your healthcare professional or pharmacist for clarification or advice.

Never take someone else's medications. Medication identification through different doses, mixing with other medications and many different names is confusing at best. Therefore, again, it is just not a good idea to share.

Children, Kids, Pets and Visitors

Kids & pets are curious; do not give them a chance to play with your equipment or medications. Do not save leftover medications (take them to the pharmacy or flush them) and do not keep your medications in the medicine cabinet in the bathroom. The bathroom is a very private place for visitors. Your children and grandchildren are very important loved ones and you should not tempt them or give them the opportunity to obtain medications that can be deadly or habit forming. Actually, this is how many illegal drugs are obtained.

Note: It is a good idea to have the Poison Control Center phone number on your refrigerator for emergencies. We all know it is in the front of the phone book, but who thinks straight in an emergency? You can add it to your medication list (which should already be on the refrigerator.

Medication Adherence (or taking your medications as prescribed)

Patients who do not take their medications as prescribed are placing themselves at great risk for death. This is a serious public health issue and there is no singular cause or solution. Many patients have more than one potentially serious illness and/or chronic illness requiring complicated daily medication schedules.

There are many causes for not taking medications as prescribed.

> Expense or cost can make it impossible to afford. Often patients will take expired medications, which can cause dangerous adverse reactions. Some will take medications not prescribed to them. These practices are also dangerous; medications come in many strengths and combinations of other medications.

Note: Never take someone else's medications.

> Forgetting to refill medications is easy to do, but can have devastating effects. Writing a reminder on the calendar or using an auto shipment program could solve this problem.

> Forgetting to take them is also easy to do, but can also have devastating effects. A daily, weekly or monthly pillbox may be the ticket (keep it out of sight in your bedroom). Try to take the medications as prescribed and at the same time of the day. This will help you develop a good habit for medicating.

> Side effects and/or adverse reactions often deter people from taking a prescribed medication. If you have a side effect, adverse reaction or develop any reason for not taking your medication, call your doctor for advice. Proper medication dosing may be the only problem, but he/she will not be able to help you if you do not communicate your concerns.

> Many people just decide not to take them or do not understand why the medications are important, how they help them, or even if they are working. Unless you have a license to practice medicine, this is a really a bad idea. In most case however; this decision not to take medications as prescribed is associated with language barriers (they just do not understand) or low literacy (they cannot read well enough to review the medication instruction sheet).

> Some patients may read an article, see something on television or get a recommendation from a friend or family member to try self-medicating (i.e. herbs, spices, animal parts etc.). This is another bad idea, without talking to your healthcare provider. Actually, you should never, NEVER start any over the counter (OTC) medication (i.e. herbs, spices, vitamins, minerals, animal parts, etc.) without talking to your doctor or pharmacist first.

> Example: Omega 3 Fatty Acids can increase the blood thinning effects of aspirin, coumadin and other prescription blood thinner medications, which could be life threatening.

Taking an extra dose because it worked so well is also a bad idea. Too much of any medication can be deadly.

Dementia (confusion or loss of memory) is also a major cause for medication problems. If the patient lives alone this is a difficult one to deal with, short of a full time medical facility.

Patient and family education, communication and medication review are essential but a personal support system (family and friends) can solve all of these problems especially if you live alone. Additionally, you and your support team must be honest with your healthcare provider. If you take too much, or did not take the medication, tell your healthcare provider. Your healthcare provider bases changes in medication schedules and doses on tests and observations when he/she sees you. If you are not taking the medications properly, for whatever reason, tell them.

Oxygen Safety

Oxygen and fire do not mix well. Well, they do, but it is not what you want to happen. Oxygen is not flammable, but it supports combustion. It will cause most any thing to burn extremely fast and drastically increase the heat from the fire. Your nasal cannula will burn like a fuse, leaving permanent scars from your nose, up both cheeks, around your ears and across your neck. These scars are not pretty.

The use of petroleum products with oxygen is not a good idea either. The oxygen can cause the petroleum to explode. You will notice that "USE NO OIL" is engraved on all oxygen fittings. This should bring up a thought; petroleum jelly is petroleum based. It will not explode; however, it may dry your lips or nose even more. There are many lubricants and moisturizers available on the market that does not contain petroleum you can choose the one you like.

If you use a liquid oxygen system, freeze burns are a real possibility. Use only the recommended procedures for filling portable liquid oxygen systems (contact your home healthcare provider for instructions; they will have a Respiratory Therapist on staff). It never hurts to think safety and have a pair of good gloves handy. Use caution not to knock over the large storage tank.

If you use oxygen cylinders or bottles at home, you must think tank safety. Oxygen bottles have up to 3500 psi inside, having the potential of being quite dangerous. To put this in perspective, imagine an exploding car tire that only takes 28 to 32 psi to inflate. If a tank were to fall over just right, the valve could be broken off and it could fly like a missile or shatter explosively. Do not leave cylinders standing up by their self; place them safely in a stand. If you travel with oxygen in your car, make sure they cannot fly around the car in an accident. Oxygen tanks are considered safe at temperatures up to 125 degrees Fahrenheit (your car can reach 140 degrees or higher). Heat causes the gas to expand which increases pressure within the cylinder. This can cause the cylinder to rupture, which will damage your car and could be deadly for anyone near the cylinder.

OSHA (Occupational Safety & Health administration) imposes the safety rules on businesses for the same reason you should enforce them at home or while traveling.

Electrical Power Loss and Storm Preparations

If you use an oxygen concentrator at home, a loss of power could be a huge problem (no oxygen). The problem is compounded if your concentrator fills portable oxygen cylinders. Make sure you have a backup supply of oxygen in case of loss of power (hurricanes, winter storms, tornados etc). Backup equipment is a very good idea, but can be costly because insurance companies do pay for them. Most home equipment suppliers have a plan to work with you.

Just as with the concentrator, if you use a nebulizer compressor at home and have power outages your nebulizer will not work. Portable nebulizers (with batteries) or a MDI (Metered Dose Inhaler) will get you through these times with out missing your medications. In an emergency, you can use your oxygen cylinder to power your nebulizer. It takes about 6lpm to run the nebulizer and this will run a tank out of oxygen in a short period. However, in an emergency you may not have another choice. An "E" Tank (the one placed in a cart and you pull around) should provide you with 8 to 10 treatments, where the smaller tanks you carry (D's, B's and M6's) supply enough oxygen for 2 to 4 nebulizer treatments.

Storm preparation is important for anyone who uses prescribed medications. Make sure you have enough on hand in advance. During a power outage, getting medications is much more difficult. It would be a good idea to have a backup power source also.

Make travel preparation with the same care and foresight as during storm preparation. It is important for anyone who has prescribed medications to make sure that they have enough on hand in advance. Traveling with oxygen can pose many difficulties. Talk to your oxygen supplier about options (travel concentrators can be checked into cargo or portable concentrators used during flights). Airlines require a doctor's letter (in some cases forms must be completed by your doctor and/or your oxygen supply company) far in advance of your date of travel, so if oxygen needs to be provided on the plane, be prepared (you can find information at www.regulations.gov).

Home Equipment

Equipment maintenance is required on most home oxygen equipment, however your oxygen concentrator conducts constant self-diagnostics and will alarm if any problems should occur. Your oxygen service provider should contact you every 1 to 3 months in order to check your equipment and provide you with replacement supplies (cannula, water bottles, extension hoses etc)

Equipment cleaning and disinfection is your responsibility. The rule of thumb is; if it is wet (nebulizers, humidifiers, etc.) they need to be cleaned at least every 3 days (72 hours) so that bacteria and stuff does not start growing in the liquids. Many home health care companies suggest cleaning nebulizers in dish liquid and disinfecting them by soaking them in a vinegar and water solution for at least 1 hour, every 24 hours. Some hospitals change out patients nebulizers each day. If you are in the hospital or a facility and they do not change them regularly, ask them to change the equipment at least every three days.

Keep power cords and oxygen supply tubing off the floor and out of walkways to avoid posing a tripping hazard. Extension cords may not be a safe idea (due to power requirements or equipment), ask your oxygen supplier for advice.

Keep oxygen extension tubing as short as possible. If they are too long, oxygen flow is reduced to your cannula and the extra lengths of hoses can be a trip hazard causing you to fall. Talk to your oxygen supply company about appropriate lengths of tubing.

Stuff that makes you sick hides

Bacteria, virus and mold can hide anywhere it can find moisture and/or food. It can be a little scary; they hide on countertops, doorknobs, bed linen, computer keyboards, cell phones, handbags, cosmetics and just about anywhere, you can think of them hiding.

Keeping things clean can save you a lot of money and reduce reoccurring illnesses. In hospitals or facilities, all healthcare workers (that is anyone who touches you or your stuff) require clean gloves.

Infections are becoming more common inside and outside of hospitals (due to a lack of cleanliness). Things like the MRSA bacteria (Methicillin Resistant Staphylococcus Aureus) and the C. diff. bacteria (Clostridium difficile). Both bacterial infections can be cured with very strong antibiotics (which are very rough on your body) however preventing the infection by cleaning is best. MRSA & C. diff can be killed by just washing your hands in warm water and soap. He FDA (Food and Drug Administration) has determined that the use of alcohol based sanitizers do not normally kill MRSA or C. diff., however it does kill many types of bacteria and viruses. In order to kill C. diff. on surfaces like countertops, keyboards, cell phones and doorknobs you will need to use a bleach solution.

Note: Showerheads and faucets are moist often-harboring bacteria or mold and should be cleaned or replaced regularly. People with weakened immune systems, recent organ replacement or surgery are at higher risk of becoming ill.

Fall Prevention

Discuss a regular exercise program with your healthcare provider. Exercise will build stronger muscles and prevent falls. Ask your healthcare provider to review your medications. Many medications can decrease your blood pressure; make you sleepy, dizzy or confused. This includes over the counter medications and herbs. Have your vision checked each year. Glaucoma, cataracts will limit your vision.

Remove trip hazards from the floor (i.e. books, shoes, dogs, cats, cloths, electrical cords, oxygen hoses etc.). Keep stairs or staircases free from clutter and install safety rails or grabs.

Try to keep frequently used items in lower cabinets to reduce the use of stepping stools.

Install safety bars at the tub, shower, toilet and sink. Use non-slip mats in bathrooms, kitchen or anywhere you have wet surfaces.

Improve lighting, as we age we need brighter lights to see well. Hang lightweight curtains or shades on windows to reduce glare.

Throw away those slippers… Wear shoes inside and out, shoes reduce the chances of slipping or twisting your ankle by supporting your feet.

Air Pollution

Air pollution is caused by almost anything that is floating in the air. The Air Quality Index (AQI) is an indicator of air quality at a specified time and location. Although the scale and numbers vary around the world, the higher the number the worse the air is for breathing.

Nitrogen dioxide, ozone, sulfur dioxide, carbon monoxide, and particulates (such as lead, smoke, dust, etc.) can make breathing difficult for people with asthma, bronchitis, emphysema and heart or lung diseases. Allergy Management requires that you pay close attention to the Air Quality Index (http://www.airnow.gov).

Bad ozone is the combination of three oxygen atoms produced by chemical reaction from pollutants here on earth and is air pollutants, which affect your lungs. Good ozone is high altitude ozone, which protects us from the ultraviolet light from the sun.

When the Air Quality Index is high (normally on hot sunny days), it is unhealthy. You should limit your exposure and physical exertion outside. Normally exercise early in the morning or in the evening is much safer for people with chronic heart or lung disease. Symptoms include; nose or throat irritation, cough, chest pain, increase in asthma symptoms, dyspnea (shortness of breath or difficulty breathing), decreased lung function and increased risk of respiratory infections or allergic reactions within the airways.

High mold and/or pollen counts can cause the same effects, so be smart and know what you are breathing.

Indoor Air Pollution

Most homes have higher levels of pollutants inside than outside in the fresh air. Chemicals from carpets, furniture, cleaning supplies and paints can affect your ability to breathe (especially if you suffer from asthma, COPD etc.). Adding live plants to the décor of your home will help to eliminate these pollutants. Having your air ducts cleaned routinely, using a better filter to your heating and air-conditioning system and using a vacuum cleaner with a HEPA filter or a good filtration system will reduce molds, pollen and dust that can make your breathing conditions worse.

Note: Ultraviolet light used in air handling systems kills germs and bacteria.

Note: Every home should have at least one smoke detector and if you have anything with that produces a flame (i.e. water heater, furnace, stove etc.) you should have a carbon monoxide detector. You should also have your basement checked for radon (colorless odorless gas, heavier than air).

Terrorism

Bioterrorism it the intentional release of biological agents (bacteria, toxins or viruses) using a bio-weapon into the air, food or water supplies. These agents can occur naturally or can be man made (or modified) to inflict widespread illness or death. Bioterrorism has been a method of warfare for thousands of years and the selection of agents is based on their ease of production, delivery and transmission. We have included some of these agents in chapter 9 (i.e. Anthrax, Smallpox, Botulism, Tularemia, Plague etc.).

A chemical terrorism attach is designed to make people sick, cause pain, cause neurologic dysfunctions (nerve transmissions) or to kill them. It is usually in the form of a dust or mist carried through the air or water in order to contaminate large areas. Inhaling, ingesting or touching the chemicals may be all it takes to make you very sick or worse.

A nuclear terrorism attack will probably not be in the form of a nuclear explosion with a big mushroom cloud. It will most likely be in the form of a "Dirty Bomb". A dirty bomb distributes radioactive materials (i.e. dust, dirt, water etc.) in order to make as many people possible sick from radiation poisoning.

Dentures

Using toothpaste and a brush to clean your dentures is not a good idea. Dentures are not as hard as your real teeth and the toothpaste can scratch them, which leave places for bacteria to hide and grow. If you aspirate (suck into your lungs) any of this bacteria you can become sick with pneumonia (see chapter 9, Aspiration Pneumonia).

If you use a CPAP (Continuous Positive Airway Pressure) or BiPAP (Bi-level Positive Airway Pressure) system for Sleep Apnea make sure you know how your mask was fitted. When fitting a mask or nasal prong appliance for patients with dentures it is much easier to maintain the proper seal with the dentures in place. If your mask was fitted with dentures in place, you may need to be refitted if you decide to take your dentures out while sleeping.

We have included this information so you can be well informed about what can happen. We hope it never happens…

Notes:

Ask your doctor about the dangers of taking too many vitamins.

Chapter 4
Legal Documents & Forms

Healthcare advanced directives or advance care planning give patients the right to decide today, how they want to handle medical decisions for their future. Every competent adult has the right to make decisions concerning healthcare including refusal of medical treatment. When a person becomes unable to make decisions due to mental or physical changes such as developing dementia, Alzheimer's disease or being in a coma, they are considered to be incapacitated. Most states have enacted legislation pertaining to health care advance directives. The law recognizes the rights of a competent adult to make advance directives instructing their healthcare provider to provide, withhold, or withdraw life-prolonging procedures. Additionally you can designate a trusted individual to make treatment decisions if you become unable to make those decisions for yourself. These advance directives can also include anatomical donations after death.

Hospitals, home health, nursing homes, and hospices are required to provide their patients with information concerning healthcare advance directives.

Advance directives are scary; full of questions about, what if's, could I's, and maybe's. However, these are not the important questions. Better questions to ask include the quality of life factors. Will mom/dad recover enough to do their favorite things (i.e. cook, walk, woodwork, got to the store, etc.)? What are their wishes?

Now let us talk about the paperwork and get a feeling for what it can do for us. Remember each state is a little different, but here is the general idea.

Advance directives completed in one state may be honored in others.

Advance directives come in four basic types, living will, healthcare surrogate designation, Do Not Resuscitate (DNR) and anatomical donation.

The living will is a written or oral statement containing what medical care you want and do not want in the event that you are incapacitated. The Living Will is a legal document. It dictates your wishes concerning medical procedures and treatment. Additionally, the choice of which medications and treatments you do or do not wish to receive can be included. Make sure your healthcare provider, attorney and significant persons in your life know where to find these papers. Additionally, your healthcare provider should file a copy and you should keep a card in your wallet or purse with information on their location.

The designation of a healthcare surrogate designates another person to be your representative in medical decision making in the event you become incapacitated. This person should understand your desires and should be the person that can make hard decisions concerning your healthcare. It is important to make sure that the person you select is willing to follow your wishes and will make these hard decisions. They should also be provided with a copy of the paperwork. Make sure your healthcare provider, attorney and significant person or persons in your life know where to find these papers. Additionally, your healthcare provider should file a copy and you should keep a card in your wallet or purse with information of where the papers are located. An alternative or addition to a healthcare surrogate is a durable power of attorney. Through a written legal document, you can name a person or persons to act on your behalf. This document can be quite important for business, financial, legal or medical matters. It is important to consider whom you select in order to prevent disagreements that may affect your wishes.

The anatomical donation form is similar to state driver's licenses for organ donation instructions. It may sound a bit gruesome, but you can help so many people with your gift, there could be no greater gift to humanity. Doctors would never cause or allow anyone to die to retrieve their organs and medical professionals take great care not to make any mistakes concerning a donor's life or death. Additionally, there are specific criteria one must meet to prove that one is no longer alive before the donation process can begin.

Which is best for you? Your healthcare provider and/or lawyer can guide you in the process. You may find that a combination of all four is the right way to go for you.

Advance directives are not required; however, by not having your desires legally documented you are placing the burden of those decisions on your spouse, children or loved ones. If there is no one, willing or present to make the decisions the court may be required to appoint a legal guardian. These people may not be aware of your intentions, religious needs and desires, or may even have their own personal agendas, which conflict with your health care demands.

Most states do not require an attorney to prepare these documents; however, they are often completed in conjunction with your last will and testament. They are legal documents and do require a witness other than relatives.

You can cancel or change the documents at any time. These changes should be written, signed and dated. Additionally, you can change your advance directives by oral statement, just destroying the old advance directive, or by writing a new one. Make sure you file a copy of the changes with your healthcare provider.

If you are terminally ill or at risk for a persistent vegetative state, you may want to consider having a pre-hospital Do Not Resuscitate Order (DNR). The DNR identifies your wishes during respiratory failure, cardiac arrest or other specific life threatening illness.

On the following pages, you will find samples of the forms we have discussed above. Remember each state may require something slightly different, but the basic idea is the same. You can also find legal guidance online (always make sure the internet site is trustworthy) or you can speak with an attorney (but in most states you do not need one).

Notes:

LIVING WILL

By declaration made this ___ day of _____. 20___, I,

_____

willfully and voluntarily make it known my desire that my dying not be artificially prolonged under the circumstances set forth below, and I do hereby declare that, if at any time I become mentally or physically incapacitated and

_____ I have an end stage condition (irreversible condition) and/or

_____ I am in a persistent vegetative state (permeate coma or brain death) and/or

_____ I have a terminal condition (injury or illness that will cause death) and/or

_____ _____
Initials

In addition, if my attending physician and other consulting physicians have determined that there is on reasonable medical probability of my recovery from this condition, I here-by direct that life-prolonging procedures be withheld or withdrawn when the administration of such procedures would serve only to prolong the process of dying.

Additionally I direct that I be permitted to die naturally with only the administration of comfort care measures or those required to alleviate pain.

I do_____ or do not _____ desire that nutrition and hydration (food & water) be withheld or withdrawn when the administration of such procedures would serve only to prolong the process of dying.

It is my intention that my physician and family as a final expression of my legal right to refuse medical/surgical treatment and to accept the consequences from such refusal honor this declaration.

In such event that my physician or facility is unable to provide the care set forth in this document, I wish to designate as by surrogate to carry out the provision of this declaration:

Name _____
Address _____
City _____ State _____
Phone _____

I understand the importance of this declaration, and I am emotionally and mentally competent to make this declaration.

Additional Instructions:

_____

_____

_____
Signature

_____   _____
Witness   Witness

_____   _____
Address   Address

_____   _____
City/State   City/State

_____   _____
Phone   Phone

DEGINATION OF HEALTH CARE SURRAGATE

By declaration made this ___ day of _____ . 20___ , I,

_____

In the event I have been determined to be incapacitated, to provide informed consent for my treatment, surgical, and/or diagnostic procedures, I wish to designate as my surrogate for health care decisions:

      Name _____

      Address _____

      City _____ State _____

             Phone _____

If my primary surrogate is unable or not willing to perform their duties, I wish to designate as my alternate surrogate:

      Name _____

      Address _____

      City _____ State _____

             Phone _____

I willfully and voluntarily make this decision and fully understand that this designation will permit my surrogate to make healthcare decision and to provide, withhold, or withdraw care with my consent on my behalf.

Additional Instructions:

_____

_____

I further affirm that this designation has not been as a condition of treatment or admission to any facility. I will notify and provide a copy of this document to the following:

Signature _____ Name _____

_____   _____
      Signature/Date   Witness

_____   _____
      Signature/Date   Witness

UNIFORM DONOR FORM

By declaration made this ___ day of _____. 20___, I,

_____

Hereby make this anatomical gift, if medically acceptable, to take effect upon my death. Indicated below are my desires:

I give:

      _____ any organs needed or parts or

      _____ only the following organs or parts or
    Initials

_____

_____

_____

Willfully and voluntarily make it know my desire that my dying not be artificially prolonged under the circumstances set forth below, and I do hereby declare that, if at any time I become mentally or physically incapacitated and _____ my body for anatomical study if needed. Limitations or special wishes:

_____

_____

_____

_____   _____
              Signature/Date   Date of Birth

_____
                               City/State

Witness	Witness
Address/City/State	Address/City/State
City/State	City/State
Phone	Phone

EMERGENCY LIST

Emergency Contacts		Contact Numbers		
Patient Name	**Medication**	**Dose**	**Taken How Often**	**MD**

Above we have created a sample of the Emergency Information you should have on your refrigerator, in your purse or wallet and given to a family member or loved one. In case of emergency (ICE) should placed in front of the name of the emergency contact in your cell phone. This will assist emergency personnel in the event you cannot communicate with them.

Note: Do not forget to update your emergency information list as things change.

Chapter 5
Death & Dying

Death will not be hurried or delayed; it comes in its own time and in its own way. An individual experience that is unique to each person and to all of his or her loved ones. As each person approaches death, he or she brings his or her own uniqueness to this last experience. This chapter is simply a guideline and your wishes or beliefs must certainly be honored. You or a loved one may experience much of the following and some things may not be experienced at all. Every end of life event is unique and everyone's experiences are purposeful.

Often it is difficult for the medical staff to determine the patient's preferences due to the anxiety of the family. For this reason the Living Will, Do Not Resuscitate Orders, and Advance Directives are so very important (see chapter 4).

During this experience, you or your loved one will progress through grieving stages; denial, anger, bargaining, depression and acceptance. Some people will bounce back and forth within the stages but until you complete the process, your outlook on your life and death will be affected in a negative manner. Many people seek professional counseling, advice from clergy, or friends and/or family.

Once your loved one has accepted the fact that they are dying you may notice some changes in them. Remember these are only guidelines and each experience is unique to you and the patient.

They may no longer want your help. They may confuse you making them comfortable with trying to save them or prolong their life. Talk to them and let them know you are only trying to make them more comfortable and that it all right to let go.

Discuss placement in hospice (normally a short-term terminal care center or home caregiver) with your healthcare provider. They are trained and experienced in dealing with the terminal patient and their family. They also can administer medications, which can make your loved one much more comfortable. Hospice can be a great benefit to all involved.

Withdrawal may develop one to three months prior to death. The first step in separation is from the world (i.e. television, radio, newspapers etc.). The next step is separation from people (i.e. friends and neighbors, distant family, children and perhaps, even those most close to them). Withdrawal can be normal, processing one's life and evaluating one's self takes a lot of space. Spending more time sleeping and less time awake becomes a normal behavior. The need for conversation pales to the importance of the loving touch of a warm hand, due to the knowledge that a life is about to be left behind.

Food is the fuel we use to keep our body going. As we prepare for death, it is normal for hunger to diminish. Liquids are still important and are normally preferred over solids. Meats are usually the first to go, then vegetables and followed by other foods. This is normal and it is okay. Speak to your physician concerning dietary support.

Confusion and disorientation may develop one to two weeks prior to death. Sleeping most of the time is normal. Talking to imaginary people, speaking of events and places that the family may not be aware of, and/or agitation may become routine behaviors. Additionally, physical changes may take place (i.e. weight loss, a decrease in blood pressure, changes in heart rate, decreases in body temperature, increased perspiration and/or changes in breathing). Skin color may also change between flushed, bluish, or pale.

Occasionally a surge of energy may occur one to two days or even hours prior to death. The patient may ask for food, to get out of bed, or want to visit with family. Quite often loved ones may not recognize this surge of energy until after their loved one's death. Do not feel bad if this surge of energy is missed, it can be very short-lived or very small events and it is but a small part of the entire process.

The way we deal with death depends on how we have lived our life. Fear, regret, unresolved relationships and unfinished business are the factors that determine how hard we will fight to postpone death. The loving touch of loved ones helps us to transition to what is to come next.

Now it is our turn, we who are left behind, for us the process of grieving denial, anger, bargaining, depression, and finally acceptance must begin. Everyone must go through this process during his or her life. Some of us are unlucky enough to repeat the process repeatedly. It never gets any easier.

A large percentage of family members and the loved ones dealing with death develop anxiety, depression and/or a very real form of the posttraumatic stress syndrome in response to their experience. There is no need to suffer; your family physician can help you manage the stress. Do not be afraid to ask for help.

All of us have suffered the loss of a loved one and there is no greater loss. Let us remember that regardless of our particular philosophy on life and death, we must agree that after one passes, his or her suffering has forever ended. When presented with a situation in which there is little chance of recovery, and the certainty of an intolerable quality of life would have existed, we must, even in the face of great personal loss and abandonment, make our decisions for our loved ones based on compassion and unconditional love.

Thank you for investing your heart felt dedication to your loved ones, family, friends, and to yourself. It has been our desire to provide the information contained within this book to you, for a very long time. We sincerely hope the information we have shared, helps you to be able to cope with the things that we must all someday experience.

Notes:

Chapter 6
Healthcare Equipment

Home Healthcare

Generally, home health care is appropriate whenever a person prefers to stay at home and needs ongoing care that family and/or friends cannot effectively provide.

"Home care" encompasses a wide range of health and social services delivered at home to recovering, disabled, chronically or terminally ill persons in need of medical, nursing, social, or therapeutic treatment and/or assistance with the essential activities of daily living. As hospital stays decrease, increasing numbers of patients require highly skilled services when they return home. Other patients are able to stay at home to begin with, receiving safe and effective care in the comfort of their own homes.

"The only dumb question is the one that we don't ask," this is a very true statement, but there are many times in our lives when we are so overwhelmed with new information that we don't even know what questions to ask. "A patient doesn't know what a patient doesn't know". This is where this book, your institutional providers and your home healthcare providers should come into play. Find out each of your caregiver's expertise and question them on the items that you would like to review with this book as your reference.

Whenever any new device is brought to you be sure that its operation and requirements are fully explained. In particular, consult with your "Oxygen Man," your Respiratory Therapist. Respiratory therapists are well-trained, licensed, health care providers whose function is to evaluate, treat, and care for patients with cardiopulmonary disorders (heart and lung disorders). Practicing under the direction of a physician, respiratory therapists assume the primary responsibility for respiratory care, therapeutic treatments and diagnostic procedures.

If you have a chronic respiratory condition there is a very good possibility that he or she will be prescribed oxygen to be used at least at night and nebulizer treatments (breathing treatments) if wheezing is present. The appropriate devices will be brought out by your home health company and should be explained thoroughly, especially the cleaning requirements of your equipment (i.e. nebulizers, CPAP, BiPAP, etc.).

The job of your home health care providers, particularly your Respiratory Therapist, is to provide the means to help you to experience the greatest possible quality of life, and "to do no harm". Certainly, the changes in your health, the new treatments, and life style changes are overwhelming, confusing, and possibly even a little terrifying. We will attempt to make the maintenance of your condition as manageable as possible.

The book will clarify many questions that you may have, and to suggest topics that you may with to discuss with your Respiratory Therapist. Your Home Health Care Company can and should be the first source of information for your respiratory concerns.

One should be very confident with the viability of their Home Health Care providers. Each company is required to have Respiratory Therapists on staff. He or she is very qualified to answer a myriad of questions, and is knowledgeable enough recommend that you should contact your physician or go to a hospital when appropriate (never wait for answers during emergencies, call 911).

If for any reason you are not satisfied with your present Home Healthcare Provider, it is your right to "fire them" and get another company to help you to oversee your health care needs (Proposed legislation may change many of your healthcare options).

Home Healthcare Patients Rights

Home care patients have the right to:

Be fully informed of all his or her rights and responsibilities by the home care agency.

Choose care providers.

Receive appropriate and professional care in accordance with physician orders.

Receive a timely response from the agency to his or her request for service.

Be admitted for service only if the agency has the ability to provide safe, professional care at the level of intensity needed.

Receive reasonable continuity of care.

Receive information necessary to give informed consent prior to the start of any treatment or procedure.

Be advised of any change in the plan of care, before the change is made.

Refuse treatment within the confines of the law and to be informed of the consequences of his or her action.

Be informed of his or her rights under state law to formulate advanced directives (see chapter 4).

Have health care providers comply with advance directives in accordance with state law requirements.

Be informed within reasonable time of anticipated termination of service or plans for transfer to another agency.

Be fully informed of agency policies and charges for services, including eligibility for third-party reimbursements.

Be referred elsewhere, if denied service solely on his or her inability to pay.

Voice grievances and suggest changes in service or staff without fear of restraint or discrimination.

A fair hearing for any individual to whom any service has been denied, reduced, or terminated, or has an account sent to collection. The fair hearing procedure shall be set forth by each agency as appropriate to the unique patient situation (i.e., funding source, level of care, diagnosis etc.).

Be informed of what to do in the event of an emergency.

Be familiar with the telephone number and hours of operation of the state's home health hot line, which receives questions and complaints about Medicare-certified and state-licensed home care agencies.

If you are not comfortable with one provider feel free to "shop for another", but do not give up on yourself and do not give up on your quest for better health.

Ask questions and be involved. After all, your health is of great concern to you and your family.

Breathing Retraining

For patients with COPD (Chronic Obstructive Pulmonary Disease) or other lung disease, dyspnea (shortness of breath or difficult breathing) instills a fear of suffocation. The resulting anxiety can be a viscous circle of progressively worsening symptoms. The fear or anxiety develops into panic that increases the heart rate, increases dyspnea, increases blood pressure and increases your rate of breathing. This increases the demand for oxygen making the dyspnea worse and the panic deepens making you worse and worse until it is a real medical emergency. The cycle can only be broken by medicating (sedating) the patient or by the patient controlling the anxiety and not allowing it to become a panic situation.

Relaxation Training is a tool you can practice prior to a panic attack in order to help minimize the physical effects of your experience. This is training that can be done almost anywhere. Practice tightening muscle groups like shoulders and arms, legs and feet. Lift your shoulders and relax them or lift your toes and relax them and repeat. These are simple and sound dumb, but they work.

Biofeedback is another relaxation technique you can practice prior to any anxiety attack. You relax by finding your happy place, thinking happy thoughts or concentrating on something that makes you feel secure. You can verify this technique is working by checking your heart and breathing rate. If you own a pulse oximetry meter (the red finger probe) you can use that instead of a watch or clock. Again, we know it may sound dumb but it will work for you.

Breathing Exercises

Diaphragmatic breathing helps the lungs expand so that they take in more air and reduce the work of breathing (WOB). The patient should lie as flat as possible face up (supine), with one hand on the abdomen and the other hand on the chest. While taking a breath in, press on the abdomen and extend the abdomen outward as far as possible during inhalations. The hand over the abdomen should move outward while the hand over the chest should not move. Once the patient has mastered the procedure while supine, they should practice doing it while sitting or standing. A good exercise program should last 20 – 30 minutes 2 or 3 time each day (BID or TID).

Pursed-lip breathing helps the patient move air out slower & breathe deeper. The patient should breathe in through the nose (4-5 seconds) and out through the mouth (6 seconds) while pressing their lips together (similar to whistling). Pursed-lip breathing decreases dyspnea (shortness of breath) and improves the ability to exercise.

These techniques are so important in enhancing your quality of life they were also addressed in chapter 1 (What Should I Know).

By practicing and utilizing breathing exercises and relaxation techniques you can greatly reduce the frequency and severity of dyspnea associated with suffocation anxiety or panic attack. Through practice, you can make their use "second nature" (you will not even notice you are doing them) and this will make a tremendous difference in a respiratory emergency. These are things you can do on your own to help yourself or practice with a loved one.

Equipment Cleaning (Disinfection)

Equipment disinfection is very important in maintaining good health. Any equipment that holds liquids (wet or damp) needs to be cleaned or replaced at least every 3 days (bacteria and mold can develop after 72 hours). Cleaning will help prevent infection due to germs growing in the damp equipment. This cleaning schedule should include nebulizers, refillable water humidification bottles (cannula, CPAP, BiPAP, etc.). Additionally, equipment that you put to your lips, put up your nose or breathe through requires cleaning. It is a good idea to ask your home healthcare provider for cleaning instructions. We recommend that nebulizers be cleaned every day. Each piece should be washed in warm water with soap (dish liquid works well) and rinsed thoroughly with warm water. As an added precaution, each piece should be soaked in a reusable mixture of 50% white vinegar and water for at least one hour. Rinse the equipment again in warm water and air-dry. The inside of the medication bowl should not be dried with a towel or paper towel due to the danger of leaving behind anything you might inhale during your next breathing treatment. This may be overkill, but it will certainly insure that one will not be breathing in germs during the next nebulizer treatment.

Some nebulizer medications are not compatible when mixed or at least not recommended to be mixed and just rinsing your equipment between medications is usually enough to remove medication residue. If you have questions, concerning mixing medications consult your pharmacist, Respiratory Therapist or healthcare provider.

Air filters on home equipment air intake covers should be cleaned or replaced routinely. Talk to your home healthcare provider for specific instructions.

Cleaning is common sense. You would not eat off a dirty plate or drink a 3-day-old glass of water. Why make yourself sick with your medical equipment, just keep it clean.

Much of the equipment used in the hospital is found in your home, although it may look a little different. Equipment can be overwhelming and confusing too, so you should not hesitate to ask questions of your healthcare provider or equipment supplier.

Note: The use of alcohol, extreme heat (boiling water or steam) or potent cleaning chemicals for disinfection of equipment may damage your expensive equipment. If in doubt, contact your home healthcare provider for advice.

Home Medical Equipment

Your medical equipment at home is designed to do the same thing as their counterparts within the hospital. The biggest difference is that you and your family will be responsible for identifying things that may go wrong. Your home healthcare professional should always be consulted if you feel any equipment is not working properly.

Medical Gases Delivery Equipment

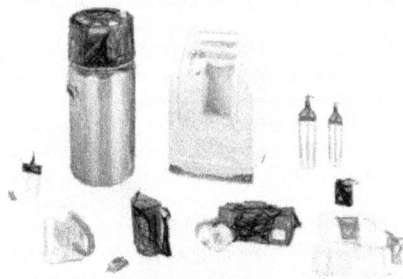

Oxygen Equipment

Oxygen can be supplied via a concentrator, liquid oxygen system (LOX), or from oxygen tanks.

Concentrators

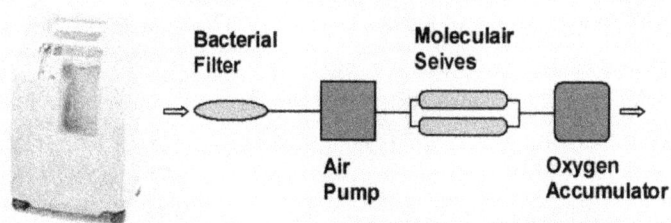

Concentrator   **How It Works**

Concentrators produce oxygen by removing other gases from the air (mostly nitrogen). Air passes through the bacterial filter, through the air pump (compressor), the air then flows into one of the molecular sieves (they alternate use, that is the switching noise that you hear during operation) and into the oxygen accumulator (which maintains a steady flow through your cannula). They have the ability to provide a flow of 1-5lpm at 95+ percentages when working properly. This is why it is important to have the concentrator checked regularly and repaired if necessary. Your home healthcare provider is qualified to render this service. It is important to remember that if the power is lost you will need a back up source of oxygen or generator to power your concentrator.

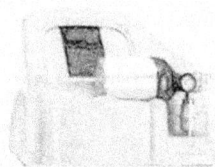

Concentrator Bottle Filling Attachment

Some concentrators can refill portable oxygen bottles or tanks of many sizes.

Portable Oxygen Concentrators (POC)

Portable Oxygen Concentrators are concentrators just like you have at home supplying 2 to 5 liters per minute (LPM) except they are smaller (weight between 1 pound and 6 pounds) and operated by battery power.

This breakthrough in technology has allowed the Federal Aviation Administration (FAA) to approve portable oxygen concentrators (POC) for domestic and international flights (you can find more information and a list of the approved models at www.regulations.gov). They are able to meet FAA requirements because they do not contain or store oxygen. Like your home concentrator, they are capable of separating oxygen from room air. The concentrator will run and charge a rechargeable internal battery (lasting about 2 - 4 hours each depending on the liter flow set) from 120vac from a wall socket, 12vdc in your car or from the DC adapter found on airliners. Be sure you check with your airline well in advance of your flight concerning any restrictions as well as any requirements from you healthcare provider and home healthcare provider. The airline may ask for a RTCA certification label, so contact your home healthcare provider with any requests for information.

Note: Since all carry on luggage must be stowed in the overhead compartments or under the seat in front of you, you may require extension tubing or a longer cannula.

Oxygen Tanks (cylinders)

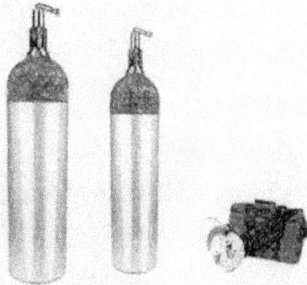

Oxygen Bottle (cylinder) with a Conservation Device

Oxygen tanks (see cylinders above) come in many sizes. Some are as tall as you are and others are small enough to be carried in a small bag. When equipped with an oxygen conservation or demand device (as we discussed above) the tanks can last for quite some time.

With any portable oxygen device (weighing 3-15lbs), it is important to know how long the cylinder will last. It will depend of the size of the tank and the liter flow you have set on the regulator. Most "on demand" or conservation devices use a battery to run its electronics and will extend the life of the tank significantly. If the battery should run down, the on demand or conservation device will free flow oxygen at the flow rate set on the regulator. This is not bad, but it will expend your oxygen supply much faster than it would if it were working properly. This is a safety feature to insure that you are not without oxygen should the battery run down or the unit fail (device failure is very rare).

Note: It is nice to have a long nasal cannula to get around, but if the hose is too long (approximately 50 feet maximum) the restrictions within the hose will reduce the output at your nose. Turning up the flow a little may not provide the flow rate you normally use. It has to do with length, internal diameter, and resistance within the tubing. Remember to always keep your hose length as short as possible and carry a portable system with you when working outside.

Let us talk about oxygen safety. Oxygen is not flammable or explosive, but it does support combustion, allowing flammable items to burn a lot hotter and faster than they normally would. Cooking over an open flame or lighting anything with an open flame while wearing oxygen is just not a good idea. The oxygen cannula will burn like a fuse and very rapidly, I might add. I have only seen it happen one time, but the scars were not pretty. Additionally, oxygen bottles are compressed oxygen, approximately 2000-3500 pounds per square inch. Bottles should always be handled with care and secured while in your vehicle. Never leave bottles standing up unless they are in a holder.

Oxygen is dry due to the manner it is supplied to you. Humidification can cheaply and easily be accomplished with a small water bottle (supplied by you home care service or hospital) placed in line at the oxygen source. Liter flow should not exceed 6lpm when using a humidification bottle (this will be discussed in detail later). Just as a note, it is not a good idea to put a humidifier on a portable tank. Should the bottle tip over, the cannula will fill with water and you will be unpleasantly surprised. If water should get into the cannula just drain it out and it will be fine. Do not to use petroleum products to relieve the dryness in your nose and lips due to oxygen use. As we all know, petroleum is flammable (it runs our cars) and oxygen supports combustion but is not itself flammable (makes a flame burn faster and hotter). When you mix the two on your skin they will react together and will irritate your skin even more. Try to use petroleum free products such as saline (salt water like tears) solutions (you can ask your pharmacist for assistance).

Your doctor will prescribe the oxygen at a certain rate of flow rate (liters per minute or lpm). If you feel you need to change the settings, call your doctor. He/she is treating you based on what has been prescribed. Additional demands for oxygen may be an indication of a changing health status and needs to be addressed. Actually, this is true for all medications, talk to your doctor.

Liquid Oxygen (LOX)

Liquid Oxygen System (LOX) with a Portable Device

Liquid oxygen system (LOX see drawing above) is a thermal insulated tank, which holds liquid oxygen (the same as the space shuttle but without the fuel) and provides oxygen flow rates up to 8lpm. Its system pressure is approximately 20-25 psi and can last up to approximately 10 days at 2lpm. The system requires no power source and fills portable liquid oxygen tanks that can last (about 8 hours at 2lpm) depending on the size of the portable tank. These portable containers are normally fitted with an on demand regulator (which use batteries) and only supply oxygen as you take a breathe (pulse of oxygen). If the batteries run out the regulator automatically switches to continuous flow at whatever liter flow is set. Your home healthcare provider is equipped to refill this system and train you in its use. The primary down side of liquid oxygen are that you must maintain a large storage reservoir (approximately 2 foot wide and 30 inches tall) in your home in order use oxygen and fill their own portable device. Additionally, LOX is extremely cold and unlike oxygen cylinder/tank will not store the liquid oxygen for up to 5 years (the liquid oxygen warms and bleeds out of the LOC portable device). Should you run low on oxygen in their LOX portable oxygen system you would be forced to return home to refill the tank.

Note: Most methods of storing oxygen have limited oxygen supply times. Ask your home medical equipment provider for guidance.

Oxygen Therapy

Now that we know where the oxygen comes from, it is time to learn how it is delivered. Below you will see the devices used, the flow rate (lpm), the percentage of oxygen supplied and explanations.

Device	Flow	FiO2 Oxygen Supplied	
Nasal Cannula	1-6 lpm	24-44%	Atmosphere is 21% oxygen, 1 lpm – 24%, 2 lpm – 28%, 3 lpm 32%, 4 lpm 36%, 5 lpm 40%, 6 lpm 44% *1
Oxymizer Nasal Cannula	1-6 lpm	28-46%	Delivers a higher oxygen level due to the reservoir device *1
Oxygen Conservation Nasal Cannula	1–6 lpm	24–44%	Oxygen delivered during inspiration only Requires batteries
Simple Mask	5-8 lpm	40-60%	Minimum flow 5 lpm to clear the carbon dioxide (CO_2) from the mask
Venturi Mask		24-55%	Set oxygen flow with manufacturer instructions
Aerosol Mask		21-100%	Set oxygen flow with manufacturer instructions/provides humidity
Partial Rebreather Mask (PRB)	8–15+ lpm	60-80%	Minimum 7 lpm with the reservoir bag full
Nonrebreather Mask (NRB)	8–15+ lpm	90-100%	Minimum 7 lpm with the reservoir bag full
OxyMask	1-15+ lpm	24-90%	No minimum or maximum flow required
Resuscitation bag AMBU Bag	15+ lpm	100%	Delivers breaths & oxygen by holding Bag mask to face and squeezing the bag

*1 There are studies that indicate you can get an increasingly higher concentration of oxygen above 6lpm to 15lpm via the nasal cannula. This flow rate however, may be far too intense to provide for patient comfort.

If you still have questions, you can talk with your doctor, home care service, http://www.fda.gov or just type "home oxygen equipment" in your favorite search engine on your computer and start the search.

Many of our patients have asked; "how do I know I am getting the prescribed amount of oxygen?" As a Respiratory Care Practitioner, we are able to verify the oxygen flow using specialized equipment. Do not get caught with a bad regulator or empty tank. The use of an OxyView™ (see the picture below) is the only inexpensive in-line flow meter we know of that is available today.

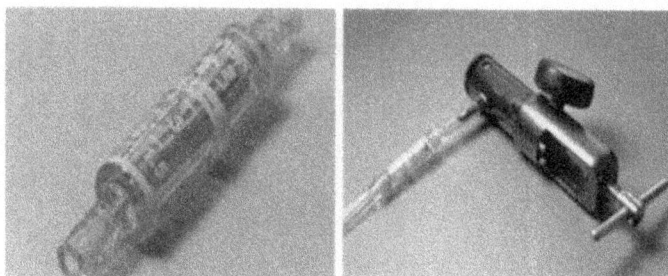

OxyView™

The OxyView is placed in line with you nasal cannula requires no batteries, works in any position and is an easy way to check your oxygen flows rates from 1 to 5 lpm at a glance. (Courtesy of Ingen Technologies Inc.)

Pulse Oximetry

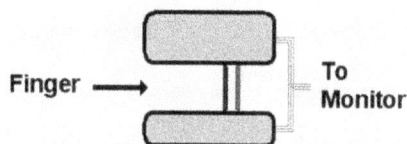

Finger Probe

Pulse oximetry (Pulsox) is the use of two frequencies of light to read the oxygen level in the blood (SpO2 or saturation) or a significant drop in oxygen level in the blood (desaturation). One is absorbed by the red blood cells caring oxygen and the other is not (or at least not as much). The difference in the absorption levels determines your blood oxygen level via pulse oximetry. The numbers displayed on your meter are oxygen level and heart rate (some systems are capable of other measurement such as breathing rate). The oxygen level displayed is in a percentage format. This means that the pulse oximetry equipment can be fooled. Here are some examples:

> People who smoke may have a higher or lower reading due to the CO (carbon monoxide) in the blood.
> Cold or shaking hands may cause a lower than actual reading.
> Poor circulation may cause a lower than actual reading.
> Polycythemia (too many red blood cells) may indicate lower level than actual.
> Anemic (too few red blood cells) patients may indicate a higher lever than actual.

Cyanide poisoning can give a high reading because it blocks oxygen from leaving the blood.

The pulse oximetry reading is representative of what may be happening within your body. If you measure your oxygen level and the monitor reads 55% oxygen and indicates, your heart rate is 22 or 200 (not normal for you) and you feel all right, try another finger or warm your hands. The heart rate should closely match your pulse rate taken on your wrist. When they are about the same, this will be your most accurate reading of oxygen level. Remember your pulse-ox is a tool, in the medical field; we try to never treat a tool, it is always better to treat the patient.

New generations of pulse oximeters have the ability to measure your Perfusion Index (PI). It is an assessment of the pulse strength or how well blood is flowing at the monitoring site (where ever you have the red finger thing attached) and displays ranges from very weak (0.02%) to very strong (20%). This measurement helps you to decide if the monitoring location is yielding an accurate oxygen saturation or measurement.

Incentive Spirometer (IS)

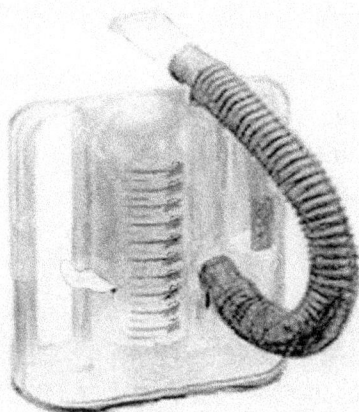

Incentive Spirometer

The incentive spirometer (or IS) is one of the most under utilized pieces in the respiratory arsenal. "It is like yoga in a can (without the stretching)". If used properly it helps to open lungs and aids in the removal of fluid in the lung tissue. Atelectasis is a (a plate like appearance on your x-ray) due to poor inflation of the lungs and Pneumonia (fluid in the lung tissue) can be stopped or prevented by using an incentive spirometer. Additionally, it strengthens your diaphragm (the muscle that makes you breath). There are no medications in the incentive spirometer; it just helps to exercises your respiratory system by indicating the strength and depth of each breath. This thing works, if used properly and routinely.

Using your incentive spirometer after surgery will help you keep your lungs healthy and clear. The incentive spirometer also will help keep your lungs active when you are recovering from a cold just as if you were performing your daily activities.

Here are the proper procedures for using your incentive spirometer:

1. Sit on the edge of your bed if possible, or sit up as far as you can in bed. Standing is also all right; however, if you were to become dizzy, you may loose your balance and fall.
2. Hold the incentive spirometer in an upright and level position.

3. Place the mouthpiece in your mouth and seal your lips tightly.
4. Breathe in slowly and as deeply as possible. If you are using the incentive spirometers for the first time watch the little thimble on the right or left side. Make sure it stays as low as possible while breathing in slowly (in the best area). You should practice breathing slowly 4-5 times while watching the little thimble and ensuring it stays low in the "best" area. You can start reading the volume by watching the large indicator cylinder (on the opposite side) and setting goals once you are sure you are using the spirometer correctly.
5. At the end of each inspiration, hold your breath as long as possible (for at least 3-5 seconds). You can exhale through the device or through your nose, (the nose is best for keeping moisture out of the unit).
6. Rest for 10-15 seconds and repeat. It is best to complete at least 10 breaths every hour or two while you are awake (do not wake up just to use the spirometer). Allow the indicator to fall to the bottom of the column between each breath.
7. The slide indicator on the side of the spirometer allows you to record your best effort. You will also find a chart for recording your progress in the package with the directions.
8. After the 10 deep breaths, practice coughing to ensure your lungs are clear. After surgery, you can splint your incision site (hold a pillow over the surgical site to reduce movement and pain).
9. Once you are on your feet again you can stop using your incentive spirometer unless otherwise instructed by your healthcare provider. Remember you can start using it again anytime you feel you are catching a cold. It is a great little tool and helps you exercise your lungs anytime.

Chest Physical Therapy

Chest Physical Therapy and Postural drainage are bronchial hygiene (lung cleaning) therapy procedures designed to help mobilize or clear secretions and improve the lungs ability to exchange oxygen. Combining percussion and vibration (PDPV) focuses a specific area of the lungs by placing the patient in the best position for moving secretions and providing percussion or vibration over the desired area of the lung. This will involve tilting the patient with their head down (if tolerated) to aid in drainage.

Postural Drainage (Positional Lung Drainage)

A procedure used in the treatment of Bronchiectasis and lung abscesses using different body positions and the head tilted down in order to help drain secretions (fluid and mucus) from the lungs. Often Chest Physiotherapy performed to loosen lung secretions.

Chest Physiotherapy (CPT) (Chest Percussion or Chest Vibration)

A procedure used in the treatment for cystic fibrosis, post coronary bypass surgery, COPD and pulmonary fibrosis. The procedure is performed by clapping on the back or chest wall with cupped hands in order to help mobilize secretions. It can also be accomplished with mechanical vibrator, a chest vibratory vest therapy (looks like a life preserver or a big blood pressure cuff), or by using an oral vibrating device (as shown below).

Note: The Therapy Vest uses an air pump to pressurize and vibrate the chest to help clear secretions (the vest looks like a large blood pressure cuff or a life preserver).

Flutter Valves

The Flutter Valve (shown above) is a device to deliver PEP (peak expiratory pressure) as well as vibration therapy (oscillatory movement) that can be used anywhere. The device consists of a mouthpiece connected to a cylinder in which a stainless steel balls or rubber lips (whoopee cushion) that are inside the valve. The patient exhales through the cylinder and causes the ball to move up and down or lips to flap during the forceful exhalation. The effect is threefold: first, to vibrate the airways to aid in the movement of mucus, to increase endobronchial pressure to help avoid air trapping (the slight backpressure helps keep your airways open) and to accelerate expiratory airflow to help move mucus out when you cough and is a form of Chest Physical Therapy (CPT). The proper use is as follows:

1. Slowly inhale beyond a normal breath (but not maximally) and hold your breath for 2 to 3 seconds.

2. Place the flutter valve in the mouth, keeping cheeks stiff while you exhale through the flutter at a fairly fast flow rate, exhaling past normal exhalation (but not maximally). You should feel your lungs vibrate as you exhale.

3. Repeat procedure for 10 breaths (it is ok to rest between breaths).

4. After you have finished your 10 breaths, initiate a cough or huff several times to help cough up any mucus.

Peak Flow

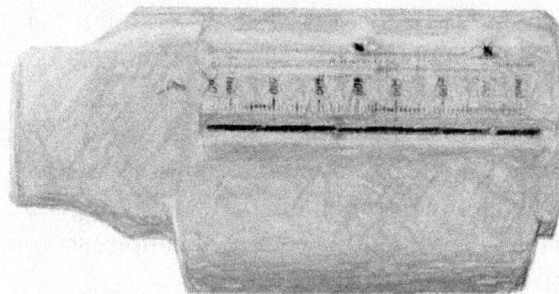

Peak Flow Meter

When you visit your doctor, they check your temperature, heart rate, respiratory rate, oxygen level and blood pressure. These are your vital signs and they are of the utmost importance in your doctor's quest to insure your good health. Elevated blood pressure, low oxygen level, very rapid or difficulty breathing are serve warning signs that something is not right and a further examination is indicated.

Asthma sufferers on the other hand, often have a difficult time gauging the severity of their asthma symptoms. This is where the peak flow meter (see picture above) comes in as an affordable and accurate way to measure your lung and airway function. It is important to note that the Forced Expiratory Volume (FEV1) is a good indicator of changes in your breathing ability; however, it may not correlate with symptom scores of asthma.

Peak Flow Test Procedure

After taking in a full breath, place the peak flow meter mouthpiece in our mouth (keeping a good seal around the mouthpiece) and exhale as rapidly and forcibly as possible. The meter does not measure how much air you exhale, it measures how fast you can move the air. The peak flow reading (measured as maximum expiratory flow rate in L/sec) is a measurement of the volume of air that you are capable of exhaling within the first one-second or less. The peak flow meter will come with a chart, which provides predicted maximal readings based on your age, gender, and height. They are not always accurate (usually due to user error) and the test should consist of the average of three procedure attempts. The peak flow meter is a valuable trend monitor. You should know what peak flow readings you are capable of achieving and understand how these numbers check or track your current asthma status.

A Pulmonary Function Test (PFT) yields a very accurate Peak Flow measurement (see chapter 11) however; the Peak Flow meter is a great tool for home or the doctor's office. The meter indicates how well your airways are working by number and color (like a stop light).

Green =   Go, good to go, or 80-100% of predicted. Your current medications are working well, so continue to follow your current plan every day.

Yellow =   Caution, watch your management efforts, or 51-79% of predicted. Consider any missed medications, pollen/mold conditions, are you catching something, or did you do something that could trigger an attack? Adjustments may be needed for your medications so, contact your healthcare provider.

Red =   Urgent, you are obstructed or at 50% of predicted, but then you already knew something was wrong. It is time to get help, now. Asthmatics tend to underestimate the severity of the signs and symptoms of an asthma attack. Asthma can be fatal and every year far too many asthmatics die needlessly due to the lack of understanding of the disease (see chapter 9).

Positive Airway Pressure (PAP)

Noninvasive ventilation is designed to provide support to ventilation through Positive Airway Pressure for patients of all ages in order to prevent the requirement for invasive (Endotracheal or Tracheal Tube intubation or artificial airways) mechanical ventilation. You may know it as CPAP (Continuous Positive Airway Pressure), Bi PAP (Bi-Level Positive Airway Pressure or continuous positive airway pressure with inhalation pressure as well as a lower exhale pressure), or APAP (Auto-Adjusting Positive Airway Pressure). They help to keep your airways open and reduce air trapping in COPD (Chronic Obstructive Pulmonary Disease) patients, reduce or prevent Atelectasis, and aide in mobilizing secretions.

High Flow Therapy (HFT) is another way of providing Positive Airway Pressure to the patient. It delivers warm, humidified air/oxygen through a special nasal cannula. Please see Chapter 1 for troubleshooting procedures for you CPAP BiPAP or APAP.

Education, fitting, titration and follow up will improve your compliance.

Adherence and Compliance for PAP

Many factors will help you be comfortable when using your relatively simple Positive Airway Pressure (PAP) therapy equipment. You must first realize and accept that you have a serious medical condition and that it is crucial to use your equipment as prescribed and use it effectively. Chances are that you will require this therapy for the rest of your life. There are many common problems and solutions listed in Chapter 1. Listed below are some ideas that will help support your needs:

> The education for you and your family is the most important. Your family can be a tremendous positive influence in your therapy. If you have special language or cultural needs requiring a translator or written materials make sure your healthcare provider or home healthcare provider understand your needs. Have your home healthcare provider demonstrate the proper procedures and you and your family demonstrate that you understand before the leave your home.

> Some people comment that the equipment is embarrassing or intrusive in the bedroom. This is why education for the whole family is so important.

> If you find the equipment confining or uncomfortable, try practicing during the day. Wear your mask while you relax (i.e. reading, watching TV, listening to music etc.) and when you take a nap. The more you wear your therapy the quicker you will become more comfortable and the better you will feel.

> Your home healthcare and healthcare provider is required to follow up with you in order to resolve any problems. This is your chance to make things right. Tell them about any problems you are experiencing (i.e. the mask, the nasal prongs, the hoses, the heater etc.). The solution may be a very simple one or you may need an equipment change. Be honest, tell them if you are using the equipment as prescribed, you are your last line of defense and you need to solve any problems.

> You have the right to choose the type of equipment you prefer to use. Have the home healthcare provider demonstrate the models they have available and if you do not find

one you will actually use contact another home health care company. Often, due to so many systems being available and the expense of having one of everything your choice may be limited with only one home healthcare company. Every one is different and your requirements may be different from others. Equipment is developed and improved all the time. If your old equipment no longer supports your needs research one that will. You can always ask your home healthcare provider for assistance.

Consider a portable system that you can take with you for travel or the camp. The equipment is small enough to carry in you luggage and the newest generation of PAP equipment is battery operated. Contact you Home Healthcare Provider or Healthcare Provider for any special requirements for airlines, high altitude or other countries (power requirement).

When fitting a mask or nasal prong appliance for patients with dentures it is much easier to maintain the proper seal with the dentures in place. If your mask was fitted with dentures in place, you may need to be refitted if you decide to take your dentures out while sleeping.

Note: Your acceptance, positive reinforcement, and compliance with you PAP equipment WILL give you an overall improvement in your quality of life through improved management of your overall health.

CPAP & BiPAP

Continuous Positive Airway Pressure (CPAP) Machine

Continuous Positive Airway Pressure (CPAP) or Positive Airway Pressure (PAP) is a treatment provided by a machine (see picture above) and mask worn at night or during times of sleep to treat obstructive or central nervous system sleep apnea, both being sleep disorders in which a person regularly stops breathing 10 seconds or longer (see chapter 11). Untreated sleep apnea can increase the chance of developing high blood pressure, congestive heart failure (CHF), heart attack, severe headaches and/or stroke. Untreated sleep apnea can also increase the risk of diabetes, the risk for work-related accidents, and driving accidents (see chapter 9).

A CPAP machine holds air pressure (from 2-25 cmH2O (centimeters of water)) in the airways, keeping tissues in the airway from collapsing when a person breathes. CPAP is the most widely used treatment for sleep apnea caused by blocked airflow (obstructive sleep apnea).

No matter what type of Positive Airway Pressure (PAP) device you use, it is a good idea to add a HEPA (High Efficiency Particulate Air) filter (a very fine filter or 0.3 micrometers) at the discharge port (where the air comes out) of the PAP machine. The filters on the back or bottom of the machine just do not filter out small particles. These extra filters are inexpensive and a really good idea (see the drawing below).

Filter

Note: If you change to a heated humidifier, you may need to change your Positive Airway Pressure (PAP) tubing in order to reduce water forming (rainout) within the tubing. Ask your home healthcare provider for guidance.

CPAP/BiPAP Masks

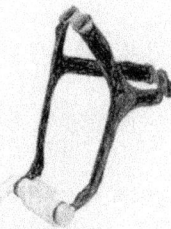

Nasal Mask   **Nasal Pillows**   **Full Face Mask**

The CPAP machine delivers air through a mask (see picture above) that covers the nose and/or mouth (the Full Face mask covers the mouth and nose). The most common type of mask used covers only the nose or uses prongs that fit inside or seal directly at the opening of the nose (nasal continuous positive airway pressure, NCPAP).

If the CPAP system requires a higher setting, which is uncomfortable, a Bi-level Continuous Positive Airway Pressure (BiPAP) may be the ticket. This machine is able to deliver the higher pressure, which is still required to relieve upper airway obstruction during inhalation; and a lower expiratory pressure to increase comfort. The types of patients that fit in this category are those with chronic lung disease, those who have chest wall deformities, and those who have a significant drop in oxygen level in the blood (desaturation or hypoxemia)

Many companies have developed an automatic CPAP, which automatically adjusts the supply pressure based on your changing requirements or your OSA (Obstructive Sleep Apnea) needs at that time.

Central apnea (neuromuscular disorders) patients require a backup rate (the system will trigger a breath) that guarantees a minimum respiratory rate during these apneic periods. This would require a different type of system and although done in the home these patients will require monitoring.

An increase of carbon dioxide (CO_2) is common in patients with untreated sleep apnea. It is important to understand that excess carbon dioxide (hypercapnia) is an important marker for the quality of ventilation. It your carbon dioxide is high, something is wrong within the cardiopulmonary system (heart and lungs). Common causes (discussed in chapter 9) are chronic obstructive pulmonary disease (COPD or Chronic Obstructive Pulmonary Disease), emphysema, asthma (during an attack), central apnea (neurological cause), and neuromuscular diseases (CSA, ALS) that cause weakness in the muscles of respiration (the diaphragm). CPAP & BiPAP are used to normalize carbon dioxide (CO_2) and oxygen delivery during these periods when breathing is stopped (apneic periods).

It is important to understand that an increased quality sleep will increase your quality of life by enabling you to increase your activities of daily living due to higher energy levels (see Sleep Apnea chapter 9).

Note: Using nasal or full-face masks can cause facial skin breakdown (sores) on the bridge of the nose or cheeks. The sores caused by having the mask placed too tightly on the face in order to get a good seal can be corrected by loosening the mask a little. Consider using the "newer" a nasal prong type instead of a mask.

Note: See chapter 1 for troubleshooting ideas for common PAP related problems.

Oral Devices

Patients with mild to moderate sleep apnea who do not require Positive Airway Pressure (PAP) therapy or patients who have not been helped by oral surgery or PAP may be candidates for a Mandibular Advancement Device (MAD), a Mandibular Advancement Splint (MAS) or a Tongue Retaining Device (designed to keep your airway open while you sleep). Historically a dentist or orthodontist were required to measure and fit the appliances that extend the lower jaw outward or hold down the tongue to maintain an open airway. Today these devices are available over the counter (OTC) using a boil and bite procedure. Ask your healthcare provider if this type of oral appliance is right for you.

There are also combination oral-nasal pillow devices that require no headgear. See the drawing below:

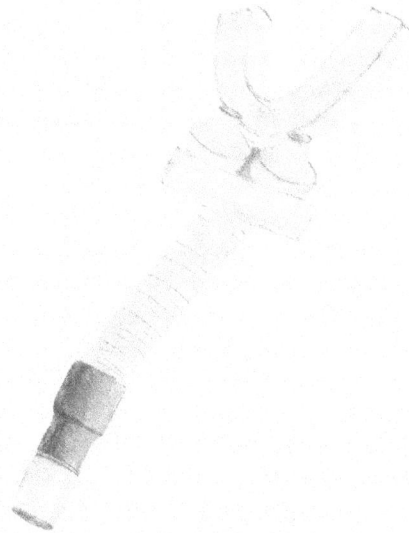

Oral-Nasal Pillow

The bite guard connected directly to the nasal pillows and the CPAP/BiPAP supply hose. This allows your bite guard to hold the nasal pillows (or prongs) in place within the nose without using any type of headgear or straps.

Nasal Pillow/Prong Device

The nasal pillows or nasal prongs look like this close-up:

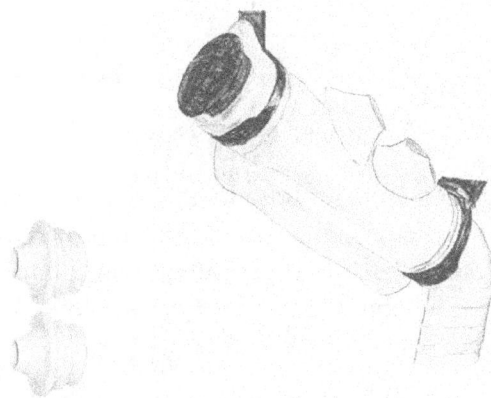

Nasal Pillows   Nasal Pillow Mask

CPAP/BiPAP and Oxygen

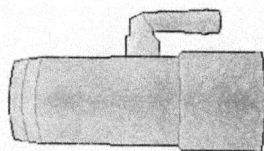

Oxygen Connection

If you should require supplementary oxygen while using a CPAP, APAP or BiPAP there is a special connection (see picture above) available to bleed oxygen into the breathing circuit. Most masks also have ports allowing you to connect an oxygen supply however; this may make the mask a bit more cumbersome due to the extra oxygen hose connected to the mask.

Positive Expiratory Pressure (PEP) Therapy

PEP Device

Low and high positive expiratory pressure therapy (see drawing above) supplements airway-clearance (mucus removal) methods. The benefit of PEP therapy is the splinting of the airways (preventing them form collapsing) which allows air to get deeper into the lungs and allowing the secretions (mucus) to be moved better by coughing. When placed inline with the nebulizer-breathing treatment it will aid in getting the medications deeper into the lungs.

Nebulizer Pump & Ultrasonic Nebulizer

Nebulizer Compressor

To use a nebulizer, you attach the nebulizer hose to an air compressor (see picture above). The compressor takes air from the environment and turns it into enough flow 5 to 6lpm) to atomize (make mist) your medications. Unlike a metered dose inhaler, which only takes a couple of minutes or less to use, a nebulizer requires you sit down and relax for 8 to 20 minutes while you inhale the medication.

Note: The newest type of nebulizer is the Vibrating Screen Nebulizer. It functions just its name implies and is reduces costs for hospitals.

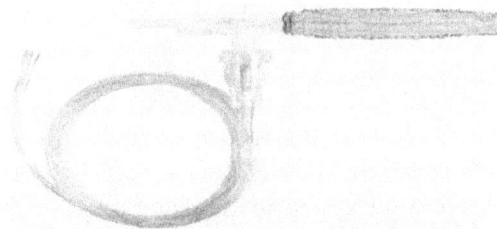

Nebulizer

Many patients and medical personnel call a Nebulizer treatment a "breathing treatment". The Nebulizer creates a mist from your liquid medications, which makes it easy and comfortable to breathe the drug deep into your lungs. If you use a Nebulizer, your doctor will prescribe the mediations in liquid form, as opposed to a Metered Dose Inhaler (MDI) or Dry Powder Inhaler (DPI).

Nebulizer Bowl

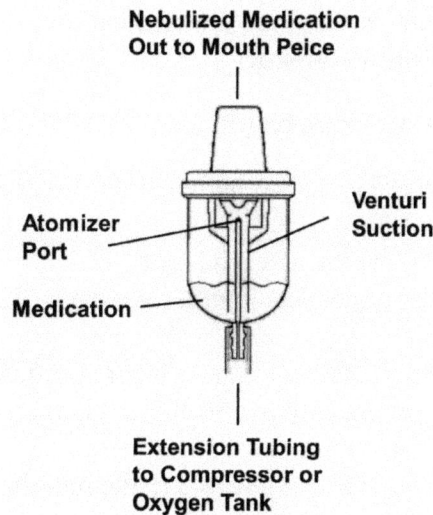

The liquid medications are placed into a small cup (see drawing above). Air from the compressor sucks up the medication using the Venturi effect (the movement of air, which sucks in the liquid medication) and atomizes (making a mist) the medication using a pinpoint stream of air and liquid. As you breathe through the mouthpiece, the mist (atomized medication) is inhaled into your lungs. You first expect it to make you cough, but the particles are so small it does not seem to bother most people.

Most compressors are small and lightweight, making them easy to use at home or away, and are compatible with most nebulizer kits. However, some nebulizers called "ultrasonic nebulizers," use sound vibrations to create the drug aerosol. These units are quieter but much more expensive and not recommended for use with some medications.

Note: Breath activated nebulizers take a little longer but less medication as opposed to continuous flow nebulizers discussed here.

Portable Nebulizer

Portable Nebulizer

A portable nebulizer can go with you wherever you go. You charge the battery or plug it into your car or an electrical outlet anywhere (outside the continental United States you will need a converter). It works the same as its big brothers and you may find them very convenient. Battery powered portable nebulizers are very convenient during power outages.

Using a Nebulizer: Instructions for Correct Use

Many medications are available as inhaled treatments because they are delivered directly into the lung and small airways. This is most beneficial in treating your lung disease or airway problems. You have a choice in how you take your medications and in what medications you actually take. Take an active interest, look up the medications, consider all treatment modalities and discuss them all with your doctor.

Assembly of the Nebulizer and Air Compressor

1. Place the compressor where it can safely reach its power source and where you can reach the ON/OFF switch.
2. Wash your hands prior to preparing each treatment.
3. Use a clean nebulizer. Routine cleaning (at least every 72 hours or 3 days) is a health imperative.
4. Measure the correct dose of medication and other solutions prescribed by your physician and add these to the nebulizer (refer to chapter 10 and medication information inserts when combining medications).
5. Connect the air tubing from the compressor to the nebulizer base making sure all connections are tight.
6. Attach a mouthpiece to the nebulizer.
7. Turn the compressor on and check the nebulizer for misting. Sit back and enjoy your treatment.

The proper way to use the nebulizer is to take a slow, deep breath, holding your breath for 3-5 seconds to allow the medicine to be delivered deep into your lungs. An aerosol mask can be used instead of the mouthpiece for those who just cannot use the mouthpiece. However, you should be involved with your medical care as much as possible, so, if you can hold the mouthpiece you will concentrate on your breathing and receive the maximum effects.

Note: It is important to remember the non-portable nebulizer requires power. You should always have a backup Metered Dose Inhaler (MDI) or portable nebulizer available for use in case of emergences or power outages. In an emergency, the standard nebulizer will operate with an oxygen tank with the flow set at 5 to 6lpm however, that will use a lot of the oxygen.

Caution: Germs grow in moist places after about 72 hours and most hospitals change nebulizers each day (if yours does not, ask for a new one). Your hand held nebulizer comes apart for cleaning. Vinegar and water work great as the final step in cleaning and removes any soap residue. You should allow each piece of the nebulizer to air-dry before reassembly. Never dry the inside of the medication bowl due to the possibility of foreign material being deposited inside and then inhaled during your next breathing treatment. Additionally, studies indicate that nebulizers only work properly for approximately 40 treatments. So change them out as needed or as directed.

Hand Held Inhalers

Inhalers come in many forms, colors, styles (as described in chapter 10), dosage levels, and number of doses (between 40 and 200). Some are Dry Powder Inhalers (DPI), CFC Metered

Dose Inhalers (MDI), HFA Metered Dose Inhalers MDI) and some are of special design. The CFC Metered Dose Inhaler (MDI) is the older style MDI using a propellant (CFC or Chlorofluorocarbons) that is not echo friendly. The newer HFC Metered Dose Inhalers (MDI) uses and echo friendly propellant (HFA or Hydroflouroalkane) for the dry powder inside and they are more expensive. The Dry Powder Inhalers (DPI) use dry powder or internal scrapings from a dry pill and your breathing in (inhalation) to move the medicine into the lungs.

One out of every two people using hand held inhalers do not use them correctly. The proper procedure is taking a deep breath, blow it out fully, and as you start a slow inhalation, administer the medication. After inhalation, the longer you can hold your breath, the more medicine you will keep in your lungs. Follow your doctor's order if additional doses are required. Normally you should wait 3 to 5 minutes (follow your doctors directions) before repeating this process. Air spacers are available when using an MDI (see drawing below) for people who have timing problems with breathing and triggering the medication. The MDI chamber or MDI spacer attaches to the MDI and allows you to be a little more flexible with inhalation.

Metered Dose Inhaler (MDI)

MDI

A CFC Metered Dose Inhalers (MDI) and HFA Metered Dose Inhalers (MDI) should be used only as directed. Never pump it more than the prescribed times because you may cause an overdose, they are expensive and you will be wasting medications. Make sure you read the instruction on the medication's package insert and air chamber prior to using one.
Note: Remember to mark, separately bag or do something to identify your "Emergency Inhaler".

The best way to check your MDI is to shake it; you will feel the medication move around in the container. It is best to shake the container prior to administering your medication dose.

Note: Reusable digital dosage counters are also available to count your MDI doses are available for around $25. These are really a good idea in order to insure you never run out of your medications.

Water is no longer recommended to be used to check your MDI for how much medication is remaining (see drawing below). Because of the fuzzy line in recognizing the differences between Dry Powder Inhalers (DPI) and a Metered Dose Inhaler which contain dry powder water should never be introduced to an inhaler. Never allow Dry Powder Inhalers (DPI) or HFA Metered Dose Inhalers (MDI) to get wet or blow into them.

Note: This method is no longer recommended by medical professionals. If this method is used on any type of dry powder inhaler (DPI), the inhaler will have to be replaced.

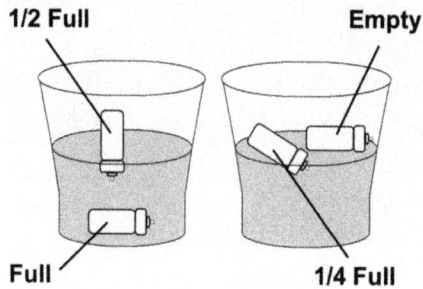

MDI Check

MDI Spacer (Air Chamber)

MDI spacers or chambers are handy for people and children who have trouble coordinating administering the medication and breathing in deeply. They are designed to hold the MDI medication for a brief period in order for you to inhale the medication properly. See the drawing below:

MDI Spacer

Dry Powder Inhaler

DPI

A Dry Powder Inhaler (DPI) is just that, dry powder. Never get them wet or blow into a dry powder inhaler, because you will blow out the medications. Never pump it more than the prescribed times because you may get overmedicated, they are expensive and you will be wasting medications. If the inhaler requires you to put a small capsule inside make sure you place it in the proper place and only press the activation button one time (or as directed). A malfunction in medication delivery or destruction of the capsule is possible if activated more times than directed. After opening a DPI their shelf life (us by date or expiration date) is very short, so make sure you read and understand all the instructions.

Note: It is a good practice to rinse your mouth out after using inhalers. Some (normally steroids) can cause thrush and some just taste bad. It is just a good idea to make it a habit.

Additional information is located on the package insert (very small print), at http://www.fda.gov or at the medication manufacture's website. It is important that you understand what you are taking, why you are taking it and the proper administration of the medicine. There are so many types of inhalers, you may find them confusing, but each one has a purpose and can really improve the quality of your life when used correctly and on time. Again, this is true with all medications.

Insufflation & Exsufflation or Cough Assist Equipment

Patients that are unable to produce a good cough could benefit from a Cough Assist Insufflators – Exsufflators (pushes air into and sucks it back out of your lungs to help produce a cough). The system has a manual and automatic settings designed to assist you in producing an effective cough. It is so very important to clear secretions from your lungs/airways and this system helps to get the mucus out. They are easy to operate and they will help you reduce the number of infections to your pulmonary system (lungs).

Note: The cough assist procedure can accomplished by hand without using mechanical equipment.

Endotracheal Tube (ET Tube)

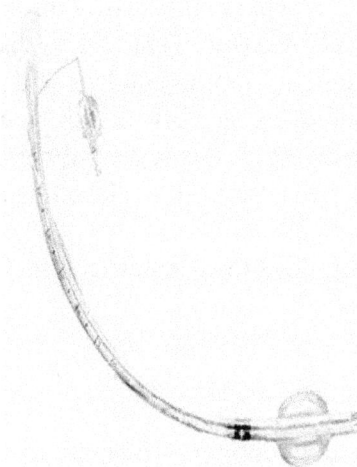

ET Tube

The Endotracheal Tube (ET Tube) is a tube about 3 ½ inches to 9 ½ inches long, which is used to secure the airway and provide a route for air to get to the lungs (when using a bag valve mask or BVM and ventilator). The proper size for an ET Tube is about the same size as the patient's little finger. Measurements are by internal diameter (called French) from 2.5 to 4.0 for infants, 3.0 to 6.5 for children and 6.5 to 9.0 for adults. Smaller ET Tubes do not have the balloon or cuff on the lung end to maintain a seal in the trachea. The proper position is just above the carina (where the trachea splits off and goes to both lungs (see chapter 7). While an ET Tube is in place, the patient cannot talk because it passes through the vocal cords and the cuff prevents air from returning through the vocal cords from the lungs. There are other specialty types of ET Tubes (independent lung ventilation, jet ventilation, double lumen etc.) which will not be discussed here.

Some ET Tubes have a suction port just above the cuff or balloon in order to remove secretions preventing them from draining into the lower airways. This helps reduce the chances of developing Ventilator Associated Pneumonia (VAP).

Note: You and your family should demand a skin and hair friendly adhesive and comfortable ET Tube holder. Cushioned, secure, flexible and easy to use ET Tube holders make you more comfortable, reduce skin breakdown and make oral care less risky (reducing contamination of the airways).

Tracheal Tube (Trach Tube)

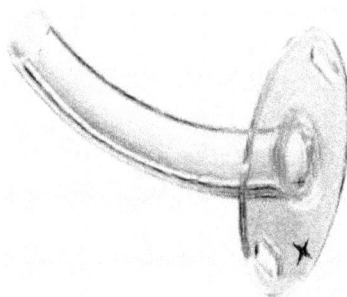

Trach Tube

The Tracheal Tube (Trach Tube) is about 3 inches to 8 inches in length depending on the type required (see drawing above). It secures the airway through a small hole in the neck providing a route for air to the lungs. Measurements are by Jackson Size 00 to 12 or 13 French to 48 French. Some have cuffs (balloons) and some do not depending on the patient's requirements. They can be attached to a Bag Valve Mask, ventilator or just used to breath through. If the tube does not have a cuff (a plastic air bubble at the end), the patient usually is able to speak with the tube in place (because it is placed below the vocal cords).

Note: ET Tubes and Trach Tubes bypass the humidification device of the body (the nose) so humidification is required.

Mechanical Ventilation

Mechanical ventilation can appear overwhelming with all the lights, noises and alarms. This section is designed for those who have a family member or loved one on a mechanical ventilator and have questions.

Control is based on Volume Control (how much air) or Pressure Control (how much pressure) in Continuous Mandatory Ventilation (CMV), Intermittent Mandatory Ventilation (IMV) or Continuous Spontaneous Ventilation (CSV) modes of ventilation.

Now let us talk about alarms. An alarm sounds and no one comes running, or maybe everyone comes running into the room. That is because your healthcare providers know the sounds of the alarms and understands what the alarms mean to the patient. Do not "lose it" when you hear an alarm sound. Instead, ask what activated the alarm. This way you will learn what alarms you should listen for and understand that the healthcare provider is on top of everything. If however, you feel they are not performing their duties responsibly, do not hesitate to notify the supervisor.

Ventilation Modes

Ventilators make all family members and friends very uncomfortable. The noises, lights, sounds and alarms can be quite intimidating. Ventilator modes are also quite complicated to the untrained eye. We will discuss just the basics in modes and why they are selected. You will find the modes in alphabetic order below:

AC or VC (Assist Control or Volume Control) provides full ventilatory support. Every breath is delivered at the same volume however, the patient can trigger additional spontaneous breaths by inhaling. AC or VC is a good choice for patients with a reduced level of contentiousness.

APRV (Airway Pressure Release Ventilation) is designed to recruit collapsed alveoli and optimize ventilation while minimizing barotraumas. Useful in treating ARDS & ALI AUTOMODE is a combination of modes (i.e. PRVC & VS, VC & VS, or PC & PS). This mode allows the patient to be removed from the ventilator faster (weaning). The vent mode changes back to full support if there is an apneic period.

CPAP (Continuous positive Airway Pressure also see CPAP & BiPAP in this chapter) When you breathe out fully, the pressure of the atmosphere and that of the lungs are the same. CPAP maintains a small pressure to keep airways open after you exhale to improve oxygenation, increase compliance (reduce stiffness of the lungs), increase FRC (see PFT) and decrease Atelectasis (airless portions of the lungs). The use continuous positive airway pressure should be used cautiously in patients with hypotension (low blood pressure) because any extra pressure in the chest cavity and reduce the return blood flow to the heart.

ECMO (Extracorporeal Membrane Oxygenation) is effective in treating newborns with acute (rapid onset) respiratory failure (see chapter 9), and in treating severe acute hypoxemic respiratory failure (i.e. ARDS, see chapter 9). ECMO involves removing the blood from the body, filtering it through an artificial lung (removes carbon dioxide (CO_2) and adds oxygen) and returning the oxygenated blood to the body.

HFV or HFO (High Frequency Ventilation or High Frequency Oscillation) is used to supply ventilation utilizing rates 60 to 2000+. Terms such as HFV, HFJV, HFPPV and HFO are all used interchangeably. The tidal volumes are very low and provide adequate ventilation while minimizing alveolar collapse by using high expiratory lung volumes.

IRV (Inverse Ratio Ventilation) is utilized to recruit alveoli and improve oxygenation using a prolonged inspiratory time (i.e. ARDS and when other methods of improving oxygenation fail. The patient will need to be sedated or paralyzed because this mode is uncomfortable

IVOX (Intravascular Membrane Oxygenation) is currently undergoing trials for use in lung protective ventilatory strategies in cases like severe respiratory failure.

LPMV) Liquid Perfluorocarbon Mechanical Ventilation is where the lungs are partially filled with fluid (perfluorodecalin) and conventional ventilators provide the mechanical ventilation.

NPV (Negative Pressure Ventilators) includes the Iron Lung and the Chest Cuirass (extrathoracic compressors or outside the chest) utilizing negative pressure via suction to the outside of the chest. This causes inspiration by making the chest to rise & expand. Ventilation is controlled by adjusting the amount of negative pressure & the inspiratory time. These ventilators are not used commonly, as modern-day ventilators utilize positive pressure. However, these ventilators are still utilized for long-term management of some chronic respiratory failure patients.

PC (Pressure Control) provides full ventilatory support at a minimum guaranteed rate that is pressure limited (stops at a present pressure level). This mode is a good choice for patients with high pressures issues and helps to avoid barotraumas.

PEEP (Positive End-Expiratory Pressure) also called CPAP (Continuous Positive Airway Pressure), (also see BiPAP). When you breathe out fully, the pressure of the atmosphere and that of the lungs are the same. PEEP maintains a small pressure to keep airways open after you exhale to improve oxygenation, increase compliance (reduce stiffness of the lungs), increase FRC (see PFT) and decrease Atelectasis (airless portions of the lungs). The use continuous positive airway pressure should be used cautiously in patients with hypotension (low blood pressure) because any extra pressure in the chest cavity and reduce the return blood flow to the heart.

PRVC (Pressure Regulated Volume Control) provides full ventilatory support at a guaranteed minimum rate and allows for spontaneous breathing. This is a pressure limited mode and its main difference from VC is the vent automatically adjusts the inspiratory pressure level based on changes in lung mechanics (how well the lung moves) during each breath.

PS or PSV (Pressure Support or Pressure Support Ventilation) is used in patients who spontaneously breathe, but still need partial ventilatory support. PS assists each spontaneous breath w/ inspiratory pressure at the dialed-in level. PS helps to increase the spontaneous breathe using pressure support in order to improve ventilation and oxygenation.

SIMV / IMV (Synchronous Intermittent Mandatory Ventilation or Intermittent Mandatory Ventilation) provides full or partial ventilatory support. SIMV with PS is a good choice for most patients and is utilized for a wide range of support (i.e. apnea to extubation). Additionally, SIMV PC+PS can provide full or partial ventilatory support in combinations best suited for each patient.

VC or AC (Volume Control or Assist Control) provides full ventilatory support. Every breath is delivered at the same volume however; the patient can trigger additional spontaneous breaths by inhaling. AC or VC are good choices for patients with a reduced level of contentiousness.

VS (Volume Support) is used with spontaneously breathing patients who still require partial ventilatory support. VS is also utilized as a weaning mode (removing patients for the ventilator) or as a backup in case of apnea. VS helps to decrease work of breathing (WOB), increase ventilation, & improve oxygenation in spontaneously breathing patients.

Ventilators are capable of many different variables and combinations of settings as discussed above, however, all they are doing is providing just the right amount of support to normalize the breathing process.

Oral hygiene is important for a ventilator patient. Ventilator Associated Pneumonia (VAP) has been associated with the bacteria that form in the patient's mouth as plaque on their teeth. Hospitals provide oral hygiene in different ways (mouth cleaning or swabbing mouthwash that kills germs). The key is to get rid of the germs, before they can be aspirated (sucked into the lungs) by the patient.

Humidification

Almost all hospitals today use a liquid oxygen system to provide the enormous quantity of oxygen needed within the hospital. When the liquid oxygen is converted into a gas, it is very dry. Humidification (or water) is added using bubble humidifiers in the patient's room. Many people do not require this humidification at lower supplemental oxygen levels (2 to 3lpm) however, if the nasal cannula dries your nose out (can cause a nosebleed) you need humidification. The bottle normally does not cost anything and only takes a minute to place in line. Just ask your healthcare provider, nurse or Respiratory Therapist for assistance. At home, a liquid oxygen system and to a lesser extent, an oxygen concentrator provides a supply of dry oxygen. Humidification can make you much more comfortable however; it will require you to replace the disposable water bottle (when empty) or clean the refillable water bottle routinely (every 3 days). When empty they stop bubbling but you still have oxygen.

Note: Special humidification equipment is required for oxygen flow rates above 6lpm. If your flow is too high, the humidification bottle may leak or burst (I have never seen one burst, but I have seen a lot of water on the floor).

Humidification Principles and Terminology

During normal breathing or respiration, air and supplemental oxygen are heated and humidified as it flows through your upper airways (the nose and mouth). This process of warming and humidifying the air is extremely efficient and important. Breathing cool, dry air/oxygen can cause damage to your lungs when delivered through an artificial airway. By the time the air reaches the carina where the trachea (wind pipe) splits and enters each lung (see Chapter 7 for anatomy) it has been heated to body temperature and saturated with water vapor. In technical terms, the air is at an absolute humidity of 44 mg H_2O per liter of gas at 37 degrees Celsius. After intubation or insertion of a tracheal tube (an artificial airway) in place to support mechanical ventilation, supplemental moisture must be added to preserve lung health.

Humidification is accomplished through many methods, but water is added (using sterile water humidifiers or by using a Heat and Moister Exchanger (HME) which catches the moisture from your exhaled breath in a filter and humidifies the air when you inhale.

Chest Tube & Suction

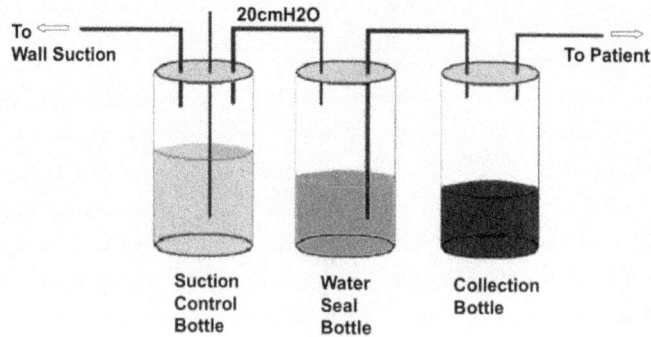

Chest Tube Suction

First hospitals do not use a setup like this for chest tube suction anymore; they have new disposable plastic systems. We drew this diagram in order for you to see how they work and for the most part all suction systems for chest tubes work the same.

The suction control bottle or chamber determines how much suction by the level of the fluid within the bottle or chamber. The water seal bottle or chamber isolates room air (the atmosphere) from the chest tube by creating a liquid seal. The collection bottle or chamber holds fluids removed from the patient via the chest tube. The collection bottle or chamber is graduated (has volumes marked on its side) in order to measure the amount of fluid removed from the patient. The plastic chest tube systems do exactly the same thing; you just cannot really look at them and be able to see how they work.

Chest Tubes sizes range from a small bore 7 French to large bore 40 French. Chest tubes are equipped with radiopaque stripes (they are on x-rays) which when placed into the pleural space (between lung and chest wall) to remove air or fluids can be easily seen in the chest x-ray. Chest tubes secured with sutures in order to prevent migration (movement) and the insertion distance recorded in order to check for tube movement.

Persistent bubbling or no bubbling at all in the water seal bottle/chamber indicates improper placement or slippage of the chest tube. Chest tubes can develop air leaks, become seated in a major fissure (tissue) or become occluded (with a blood clot or tissue), which will reduce their effectiveness.

Prior to removal, the most common practice is to check for leaks in a 4-hour clamp test. If the condition requiring the chest tube returns, it will be unclamp immediately and the cause will be determined. If removed 48 hours after leaks are no longer present, there is a 0% of recurrence.

Surgical Masks

The use of square surgical masks is better than doing nothing about breathing in germs, dust and allergens. Wearing the mask correctly and ensuring the right fit is crucial to the mask working properly. If air leaks between your face and the mask, it is not working.

Masks come in small, medium and large (most people wear a medium). If you are planning to run out and buy the N-95 style mask, you need to learn how to put the surgical mask on (read the instructions in the box or consult the internet) and you should be fit tested for the correct size (as we said most people wear a medium or a regular size). If the mask does not fit or put on correctly it will not work.

If you are immune system compromised or just worried about what you breathe in when you are around people, we recommend a surgical N-95 Respirator mask (like the HEPA Filter) which filters the air you breathe to help protect you from microorganisms including bacteria and many viruses. The N-95 is available as industrial or non-medical grade (for about 60 cents each) available just about anywhere (i.e. home improvement stores, department stores etc.) and surgical grade which is available at pharmacies and medical supply stores (for about $1.80 each).

The N-95 is disposable and cannot be cleaned or washed. It can be reused if it is clean and undamaged but remember why you bought the good mask in the first place. If the mask does not fit or is put on incorrectly, it will not protect you. People you must follow the directions.

So think and be well.

Notes:

Chapter 7
Cardiopulmonary Anatomy

It is important to emphasize that possessing even a basic knowledge of anatomy (anatomic structures) is essential to understanding how the bodily system's work. We will start at the head and work our way down through heart, lungs and body. We will start the explanations a basic level, but will include the identification of specific structures and their functions within breathing (respiratory) and heart (coronary) systems. At the end of this chapter we will put it all together as it functions to supply oxygen to your body through inhalation (breathing in), removal of carbon dioxide (CO_2) as you exhale (breathing out) and pumping blood around your body. The detailed information at the end of this chapter is in more in depth and provided in order to allow you to research individual pertinent areas of interest.

Pulmonary System (Breathing)

When we breathe in (inhale), we are breathing air that is a mixture of gases and has about 21% oxygen (the higher the altitude the lower the oxygen percentage). We breathe 12-20 times each minute or about 6-10 million times each year and most of us do not realize we are doing it. We say most of us because people with lung disease or disorders (i.e. COPD, asthma or others) often have to work very hard to catch their breath. In the following chapters we sill discuss this scary symptom of lung and heart disease.

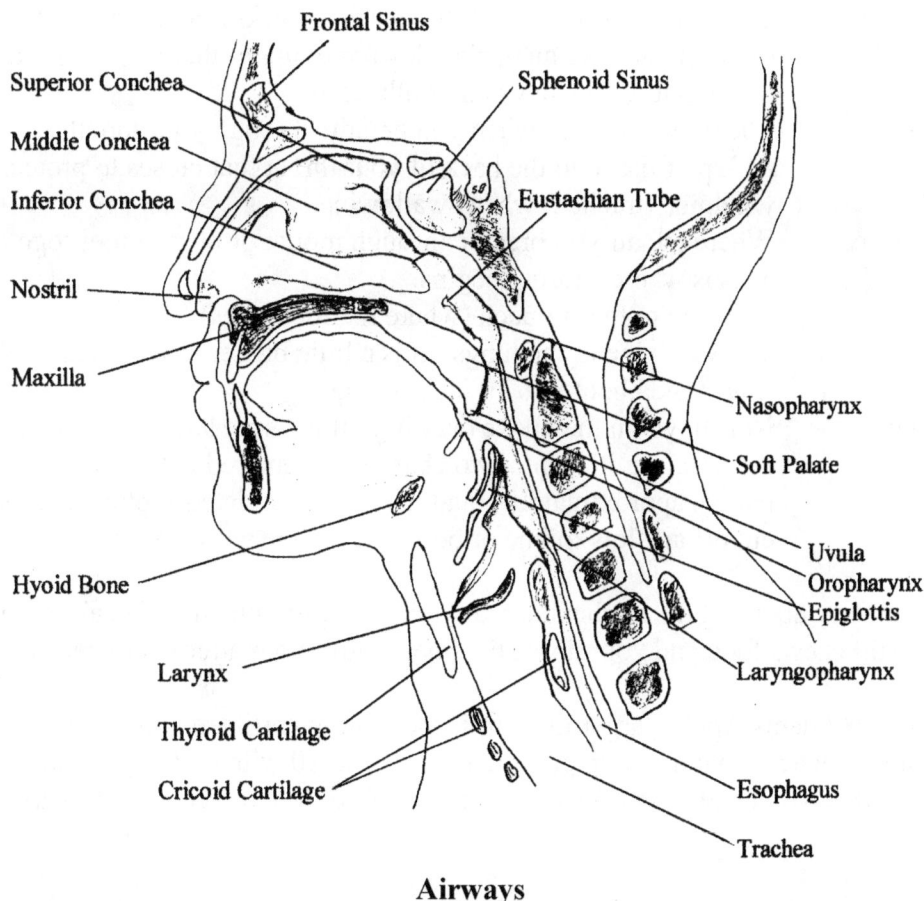

Airways

The deeper you go from your mouth into the lungs, the smaller everything gets. Now let us get started with what things do (the anatomy).

Below you will find a list of what the different parts of the airway do for you. Do not be too concerned with all the medical names; focus on what the parts does and what it means to you.

Maxilla	Bones on the top of the mouth.
Concha/Turbinate	The nasal passages (top, middle and lower) which warms and moisturizes (humidifies) the air during breathing (inhalation).
Frontal Sinus	Produces mucus and acts as a resonating chamber (like a box around your stereo speakers) for production of speech or sound.
Sphenoid Sinus	Produces mucus and acts as a resonating chamber (like a box around your stereo speakers) for production of speech or sound.
Eustachian Tube	An air tube from the nasal cavity (nasopharynx) to the middle ear in order to equalize pressure (like when your ear pops on a plane).
Note:	In small children this tube is often compressed or crimped causing frequent ear infections. Infections will slow or stop as the child's head grows into an adult shape.
Nasopharynx	The nasal passages or the route air takes when you breathe through your nose.
Soft Palate	The soft tissue on the top of your throat that closes the path to the nose when swallowing.
Uvula	That dangly peace of soft tissue (fat and skin) at the back of your throat (looks like an upside down mountain) that helps soft palate close the path to the nose when swallowing.
Oropharynx	The route air takes when you breathe through your mouth.
Epiglottis	A flap of tissue in the back of your throat that closes to protect the windpipe (trachea) when swallowing.
Laryngopharynx	Where the air you breathe through mouth or nose comes together. This is where snoring begins.
Esophagus	The tube to the stomach (where the food goes).
Trachea	The windpipe to the lungs. It is a little bigger around that you're little finger and about 7 to 10 inches long.
Nasal Septum	Not shown in the above drawing. It is tissue and bone that separates the nasal cavities into two chambers (centered between nostrils or nares), approximately equal in size. A deviated septum can reduce or cut off air flow to one or both sides to the nasal cavity.

Again, don't worry about all the medical names your windpipe still supplies air to your lungs, your mouth still chews food and your nose still smells stuff, no matter what their names.

Just behind your Adams Apple (larger in men) is the thyroid cartilage which sets on top of your windpipe (trachea). Your windpipe (trachea) is armored with cartilage to prevent it from closing when you bend your neck or move your head. Let us see how it looks.

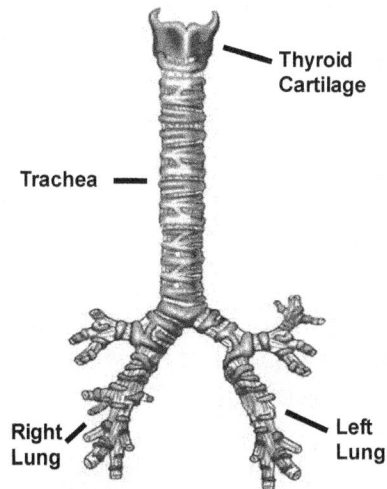

Tracheobronchial Tree

Thyroid Cartilage	Is the armor that keeps the airway open and protects the windpipe when bending.
Trachea	The windpipe to the lungs.
Right Main Stem	Where the windpipe (trachea) branches off becoming the Bronchi to the right lung.
Left Main Stem	Where the windpipe (trachea) branches off becoming the Bronchi to the Left lung.

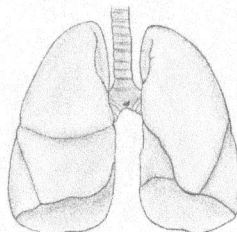

Lungs

You have a right lung and left lung (see drawing below). Each lung is divided into lobes (two on left and three on the right) and then into smaller segments. These divisions get smaller and are discussed later in this chapter in order keep things clear and simple.

Alveoli

At the end of the smallest of the windpipes (airways), you will find the alveolar sacs (see figure above). These things are tiny and it is where the exchange of oxygen (fresh air) and carbon dioxide (waste gas) takes place. Freshly oxygenated arterial blood (bright red) and returning used venous blood (dark red) encircle the alveolar sacs where the gas exchange (removing carbon dioxide and replacing it with oxygen) takes place. The red blood cells are like magnets (have an affinity) for holding oxygen and carbon dioxide (CO_2) and carrying it around the body. As you can guess, it takes many red blood cells to supply oxygen and remove the carbon dioxide for your whole body.

Note: This supply and waste function is what your doctor is checking when he/she orders an Arterial Blood Gas (ABG). It determines how well the heart, lungs and body are working.

Cardiovascular System (the Heart)

Now that we have a renewed the supply of fresh oxygen in our blood, it is time to do something useful with it, like send it around your body (to oxygenate and feed our cells). Let us discuss the heart; it is our own personal pump.

The heart is an electromechanical pump (about the size of your fist) which means it uses small amounts of electricity to make the muscles move (contract and relax) in order to pump the blood. These electrical pulses are what your healthcare provider look for in an EKG. Your heart has four chambers two on the left and two on the right. Heart valves (or check valves, see drawing below) keep the blood from flowing or leaking in the wrong direction.

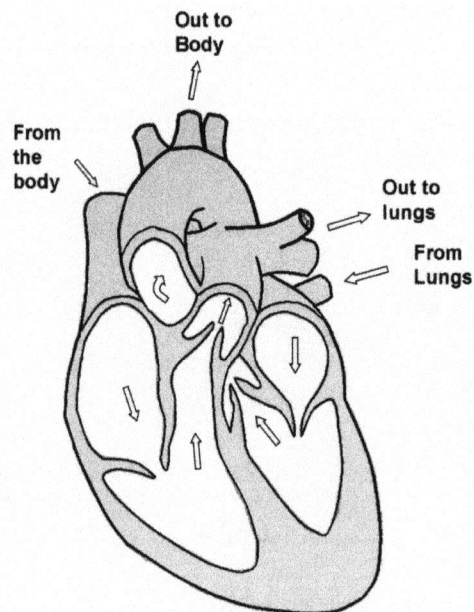

Heart Valves

There are a lot of names and functions that really are not important to you unless you have a disease or defect of the heart. Therefore, we will discuss this stuff in more detail at the end of the chapter.

The basic flow is; the heart pumps blood from the lungs to the left side of the heart where it is pumped out to the body. The blood returns to the right side of the heart where it is then sent to the lungs for oxygen. This cycle or a heartbeat is repeated 60-100 times per minute (heartbeats) or about 30-35 million times per year.

Arteries carry the bright red blood (oxygenated blood) throughout the body delivering food and oxygen and veins return the dark red blood (venous or used blood) back to the heart.

Now let us see how the heart gets its oxygen and food. Your heart is made mostly of muscle tissue and needs a lot of oxygen in order to work properly. Everyone knows someone who has had a heart attack. Heart attacks occur when blood flow the heart is reduced or stopped. This is ischemia or the lack of blood flow to the heart (or any organ). The four coronary arteries (see drawing below) supply the heart with the blood while the herd is resting (or not beating).

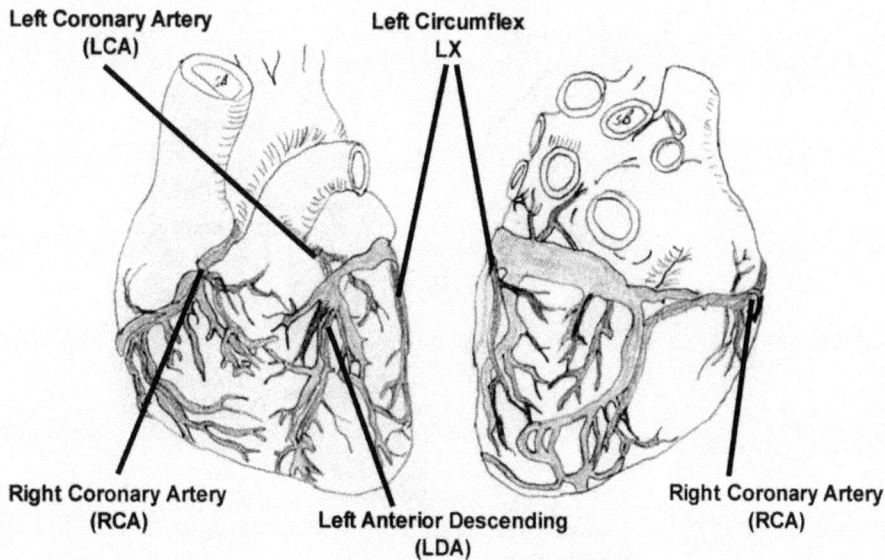

Blood Supply to the Heart

If a coronary artery were to become blocked with plaque (like cholesterol buildup) or a clot (blood clot or tissue which has broken loses) your heart muscles will start dying and cause pain. The symptoms in men and women will vary. Men's symptoms include nausea, sweating, pain in the chest, jaw, arm or back depending on where the blockage or blockages are located. Women have symptoms include fatigue, depression and jaw pain (they do not normally experience chest pain). It is important to note that the effect will vary from person to person. Do not ever think the pain will go away (you can fool yourself, but you cannot fool your body), get immediate help and do not wait. As we say in the medical field, **"TIME IS MUSCLE"** or the longer you wait, the greater the damage (heart muscle killed).

Much of the pain or discomfort during a heart attack depends greatly on where the blockage is located. These symptoms will usually differ between men and women however; you need to seek medical help immediately. Often the first heart attack is a minor one and the symptoms may subside, but do not be fooled, the symptoms will be back and you may not be so luck during the next one. Seeking help is not a waste of time and it is not embarrassing enough to risk your life.

Anatomy (a more in-depth look)

Most people reading this book now have a basic understanding of the anatomy and can move on to Patient Assessment (chapter 8). For those who need a more in-depth understanding or those who are researching something specific will want to read the remainder of this chapter. The information from here to the end of the chapter is in much more depth but is still basic. It has been included to support patients that require additional information. You may want to scan over this information because although it is basic information it can become challenging.

Airways

We will start at the epiglottis (at the back of your throat) and move down toward the lungs.

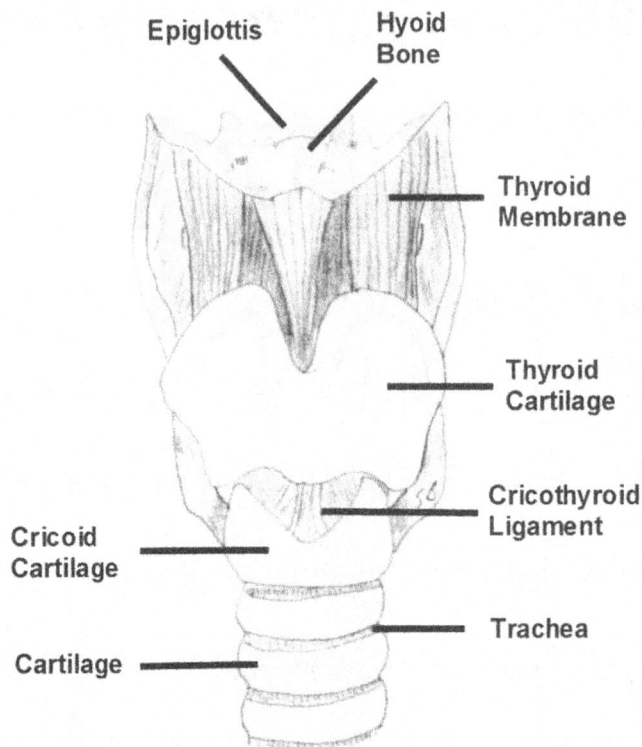

Laryngeal Cartilage

Hyoid Bone	A horseshoe shaped bone that suspends the upper portion of the thyroid cartilage by the Thyrohyoid membrane. It is the only bone in the body that is not fixed in place.
Thyrohyoid Membrane	Tissue that hold and supports the thyroid cartilage.
Thyroid Cartilage	Is the armor that protects and supports the airway.
Cricothyroid Ligament	Connects the Thyroid Cartilage and the Cricoid Cartilage.
Cricoid Cartilage	Is the armor that protects and supports the airway.

The windpipe (trachea) begins just below the Adams Apple (thyroid cartilage). As we move down the airways things get smaller and smaller. The diameter of the windpipe (trachea) is about 1 inch and the length is about 6 inches. Where the trachea split (the carina), going to each lung you will notice the Right Main Stem is a little larger than the Left. This is because the Right Lung has three lobes and the Left has only two.

Note: In anatomy and x-rays, you are looking at the patient so left and right can get confusing.

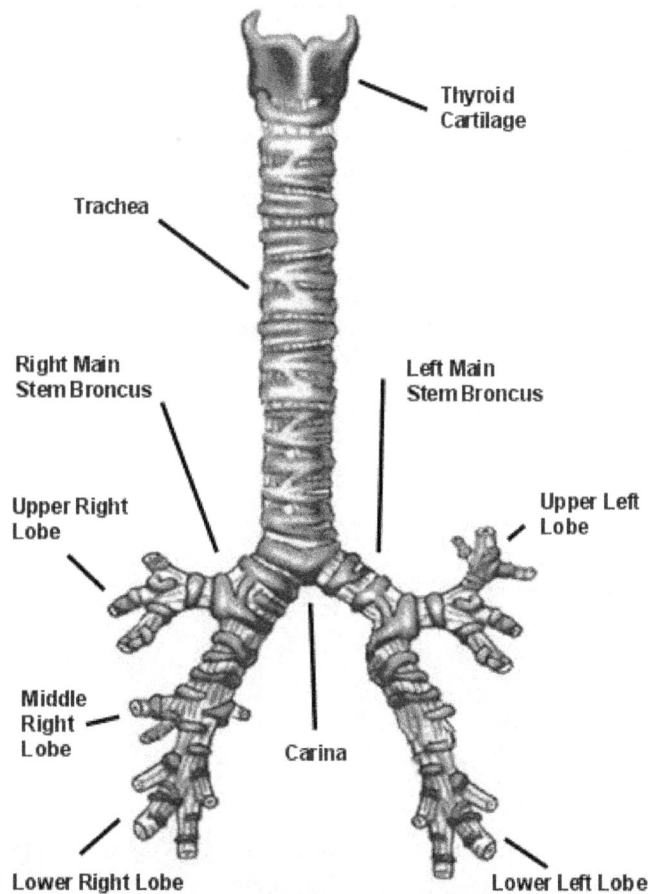

Tracheobronchial Tree

Thyroid Cartilage	Is the armor that keeps the airway open and protects the windpipe bending.
Trachea	The windpipe to the lungs. The cartilage bands add support and reinforce the airway (like armor).
Right Main Stem	Where the windpipe (trachea) branches off (at the carina) becoming the Bronchi to the right lung.
Left Main Stem	Where the windpipe (trachea) branches off (at the carina) becoming the Bronchi to the left lung.
Right and Left Lobes	Both sides of the lungs have smaller windpipes (Lobar Bronchi) inside the lungs.

Lungs

The lungs seem simple the right lung having three lobes and the left having two lobes, however, there is a lot more to them. Each lung is divided into many bronchopulmonary segments and again ever further. These divisions are the oblique and horizontal fissures. The right lung contains the upper or superior lobe (containing the apical, posterior and anterior segments), the middle lobe (contains the lateral and medial segments), and the inferior or lower lobe (contains the superior, anterior basal, lateral basal and posterior basal segments). The left lung contains the superior or upper lobe (contains apical posterior, anterior, superior lingual, inferior lingual) and the inferior or lower lobe (contains the superior, lateral basal and posterior basal segments).

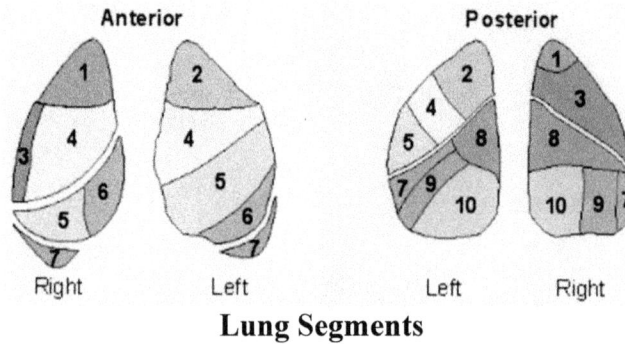

Lung Segments

	Right Lung		Left Lung
Superior Lobe	1 Apical	Superior Lobe	2 Apical Posterior
	3 Posterior		4 Anterior
	4 Anterior		5 Superior Lingula
Middle Lobe	5 Lateral		
	6 Medial		
Inferior Lobe	8 Superior	Inferior Lobe	6 Inferior Lingula
	7 Anterior Basal		8 Superior
	9 Lateral Basal		9 Lateral Basal
	10 Posterior Basal		10 Posterior Basal

When you take a breath, the oxygen in the air (about 21% oxygen) enters your mouth and/or nose flowing through the Nasopharynx and/or Oropharynx, combining at the Laryngopharynx. The Epiglottis opens allowing air to flow down through the Trachea to the Right and Left Main Stem Bronchi, and to the Lobar Bronchi (lobes of the lungs). Now things are starting to get narrow as the air enters the segmental Bronchi to the Sub segmental Bronchi, to the Terminal Bronchioles and down to the Alveolar Sacs where the gas exchange takes place. When you exhale, it all goes in reverse order (except for the epiglottis, it opens) and that is one full breath, which we normally repeat 12-20 times each minute. Any faster than 20 breaths per minute is rapid breathing (Tachypnea) and any slower than 12 breaths per minute is slow breathing (Bradypnea).

Heart

Your heart or your personal pump plays a major role in supplying food and oxygen to the cells throughout your body. We should begin with identifying what things are and what they do in your circulatory system.

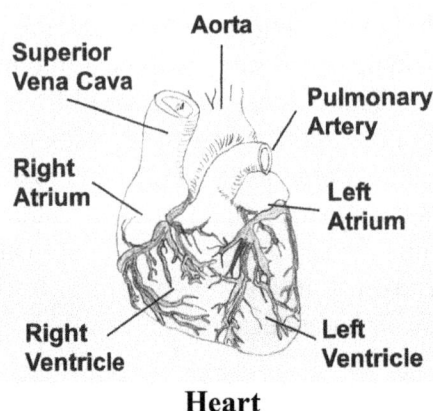

Heart

Superior (upper) Vena Cava	This large vein returns blood to the heart (Right Atrium) from the upper part of the body.
Inferior (lower) Vena Cava (not shown)	This large returns blood to the heart (Right Atrium) from the lower part of the body.
Right Atrium	Receives returned used blood (venous or dark red blood) which has higher levels for carbon dioxide and pre-fills the Right Ventricle (like taking a deep breath before blowing up a balloon).
Right Ventricle	After being pre-filled by the Right Atrium it pumps venous blood (by contracting) to the lungs to be refreshed.
Left Atrium	Receives oxygenated blood from the lungs and pre-fills the Left Ventricle (like taking a deep breath before blowing up a balloon).
Left Ventricle	After being pre-filled by the Left Atrium it pumps fresh arterial blood (oxygenated) to the body.
Aorta	Output of the heart to the body (this is the Aortic Notch which can be seen on chest X-rays).

As the blood travels away from the heart the arteries get smaller and become the Arterioles and getting even smaller yet become the capillaries. The capillary bed is where exchange of gases takes place (oxygen goes to the cells and carbon dioxide (CO_2) goes back into the venous blood). The venous side of the capillary bed (oxygen poor and carbon dioxide rich blood) expands from capillaries into Venules and again into Veins. The blood then returns to the heart via the Vena Cavas to the Right Atrium.

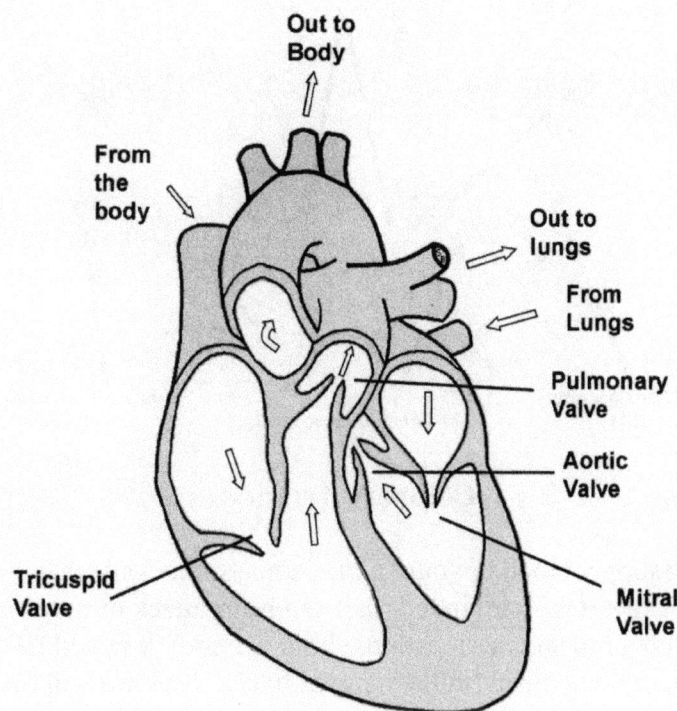

Heart Valves

Your heart has four check valves (valves that allow flow in only one direction) that control the blood flow through the heart preventing regurgitation (we have not included the coronary sinus and inferior vena cava valves). These valves are the Aortic and Pulmonic (or

Pulmonary) Valves (which are Semilunar valves), the Mitral Valve and Tricuspid Valve (which are Atrioventricular valves), see the drawing above.

The blood in the Right Atrium flows through the tricuspid valve (three sided valve) and fills the Right Ventricle and is pumped through the Pulmonic (or pulmonary) Valve into the lungs to have the carbon dioxide (CO_2) removed and pick up oxygen. Finally the blood flows back to the Left Atrium through the mitral valve (two sided valve or bicuspid valve) into the Left ventricle. This cycle completes 60-100 times per minute.

To review the blood flow, the heart pumps oxygenated blood from the Left Ventricle through the aortic valve into the Aorta. The arteries get smaller and become Arterioles and getting even smaller yet become the capillaries. The capillary bed is where the body gets its oxygen and gets rid of its carbon dioxide. The Venous side (oxygen poor and carbon dioxide rich blood) of the capillary bed expands from capillaries into Venules and again into Veins. The blood then returns to the heart via the Vena Cavas to the Right Atrium. The blood in the Right Atrium flows through the tricuspid valve (three sided valve) and fills the Right Ventricle and is pumped through the Pulmonic (or pulmonary) Valve into the lungs to have the carbon dioxide removed and pick up oxygen. Finally the blood flows back to the Left Atrium through the mitral valve (two sided valve or bicuspid valve) into the Left ventricle.

The heart muscle oxygenation (gets fresh oxygenated blood) in the process coronary circulation and it takes place during the hearts relaxation period (or diastole). This sounds odd but the heart is working too hard to allow blood flow any other time.

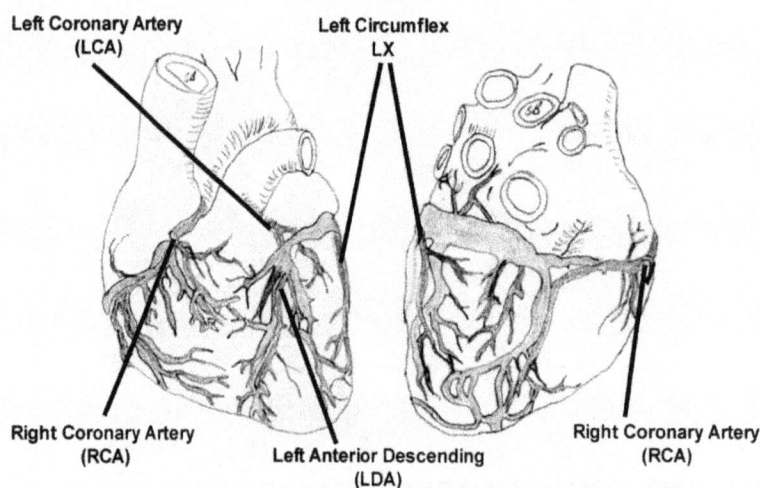

Coronary Arteries

The coronary arteries supply blood to your heart. A major blockage (ischemia or reducing the blood supply to the heart muscle) here is called a heart attack or myocardial infarction (MI) and if it should occur in another location within the body it would be called a stroke (in the brain) or a pulmonary embolism (in the lungs). In chapter 9 we will discuss diseases, disorders and in chapter 11 we will discuss corrective or surgical interventions.

Supplies blood to:

Left Coronary Artery	The left ventricle, septum, SA node, HIS bundle, both right and left bundle branches, anterior and posterior hemibundles.

| Right Coronary Artery | The right atrium, right ventricle, SA and AV Nodes, the proximal HIS bundle and posterior hemibundle. |

Note: Anterior is the front and posterior is the back. Proximal means close to whatever part that is being discussed and distal means away from whatever part that is being discussed. All these branches, nodes and bundles (electrical pathways for the heart) will be discussed next.

As we have discussed the heart is an electromechanically (electrical pulses cause movement in the heart muscle) operated pump, similar to one in your car or home. Let us discuss the amazing electrical pathways that make it function (see drawing below).

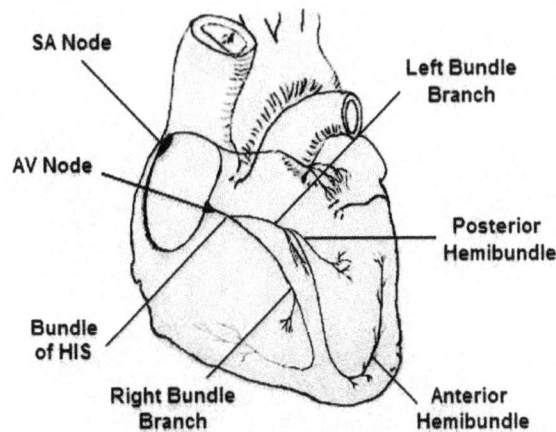

Heart Electrical Paths

The design of the heart is outstanding and quite complicated, so we will discuss the basics as it applies to its operation.

SA node	The Sinoatrial Node has a single muscle cell (the pacemaker) in the right atrium of the heart that starts the pumping sequence.
AV node	The Atrioventricular Node located in the lower half of the right Interatrial Septum (separates the right and left atria's) slightly delays the conduction of the pulse sent to it to allow the Right Atrium to pre-fill the Right Ventricle for maximum pumping ability (like priming a pump).
Bundle of HIS	Times and conducts the signal to the Right and Left Ventricles so the muscles contract in sequence.
Right Left Bundle Branch	Ensures the heart muscles contract at the proper time. This ensures effective pumping.
Purkinji Fibers	Fibers attached to the muscle to cause them to react (constrict) in order to actually pump blood.
Posterior Hemibundle and Anterior Hemibundle	Are the front and back conduction pathways.

Since the heart uses electric signals or pulses, they are monitored in respect to time and signal strength (amplitude). This display, signal verses time, is what the doctor calls an Electrocardiogram (EKG) we will discuss it in further detail in chapter 11. When you have a heart attack, these singles will change (in some cases it may take over 24 hours to change). In addition, laboratory blood work can detect a recent heart attack by looking for specific signs (enzymes) in your blood that the heart only releases during the heart attack.

It is important to note that even though the electrical pulses of the heart are present on the EKG, the heart may actually not be pumping or pumping correctly (we will discuss this too in chapter 8 & 11).

If you should have a blockage in the blood supply to the heart, it can cause an acute (rapid onset) myocardial infarction (heart attack, see chapter 9). Depending on where the blockage is located (see drawing below), side (lateral), front (anterior), bottom (inferior), or back (posterior) the symptomatic (symptoms of what is wrong) pain (in chest, back, jaw or arm etc) and effects will vary.

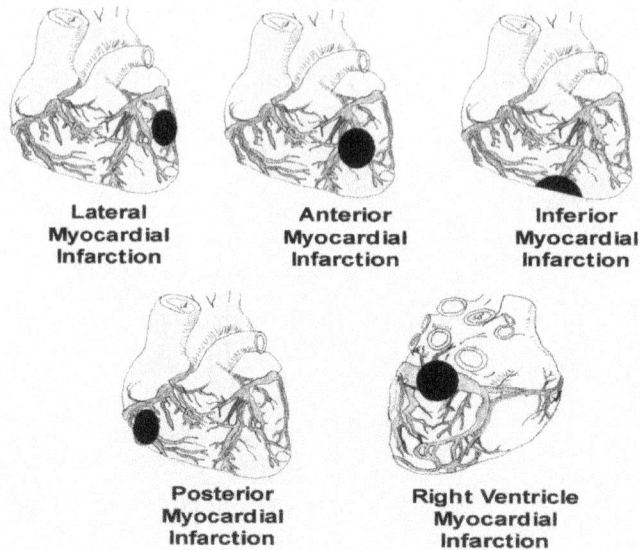

Lateral Myocardial Infarction   **Anterior Myocardial Infarction**   **Inferior Myocardial Infarction**

Posterior Myocardial Infarction   **Right Ventricle Myocardial Infarction**

Blockages Causing Heart Attacks (MI)

Do not ever think the pain will go away (you can fool yourself, but you cannot fool your body), get immediate help. The symptoms in men and women will vary. Men's symptoms include nausea, sweating, chest pain, jaw pain, arm or back pain depending on where the blockage or blockages are located. Women's symptoms include fatigue, depression and jaw pain (they do not normally experience chest pain however they can). In addition a feeling of impending doom (something bad is about to happen) and shortness of breath (dyspnea) are tell tail signs.

Remember, **"TIME IS MUSCLE"**. If you try to ignore the signs and if you do not ask for help, no one knows you are dying. That rather makes it your fault now does it not?

Note: Everyone should learn CPR (Cardiopulmonary Resuscitations), but if you have a loved one with heart problems, a pool, kids, children, hotdogs, hard candy, toys, allergies, a job, or who is still breathing CPR is A MUST (see chapter 1). Get the picture? Get er done.

Notes:

Notes:

Chapter 8
Patient Assessment

Many tests, observations, signs and symptoms are used to determine what is wrong, its cause and the best treatments to correct the problem. We will start with the basics and work our way up to the good stuff and as in chapter 7 will close the chapter with detailed assessment and testing procedures.

Chart Review (Medical Record)

We will review your healthcare chart in order to see what has been done for you and how these actions have helped or hindered you getting well.

Medical History Review

It is important for your healthcare professional to review diagnosis, symptoms, medical history, prescriptions, order changes, medicine administration report (MAR), intake/output (water/urine), and progress notes prior to entering your room. These factors often offer valuable insight into how they can better care for you. Additionally, changes in your vital signs which include temperature, pulse, blood pressure, laboratory values, X-rays, blood oxygen level (Sat or SpO2), and respiratory rate may hold valuable clues. In some cases, the clues lay in what has not worked to support your improvement. This is what we call a chart review. Before you see a healthcare professional the review is done and will help us decide, what changes (if any) are needed to support your care.

Bedside observations and interview

When we enter your room, we scan everything in the room including you. Our first impression of you and your room can tell us much (we are not looking for that extra ice cream container). Nonverbal communications (i.e. facial expressions, body movement, and patient interest) can be very important to our assessment however, we need you to talk to us.

The interview	Your team needs to know what you think, how you feel and if you have any concerns or ideas about your treatment plan. Your speech patterns are important and they can tell us a lot; can you speak without becoming short of breath and how many words do you speak between breaths.

Note: If you do not speak up and stay involved in your care, any lack of recovery is partly your fault.

Physical Assessment

The assessments include how you look, feel and sound. These are the basics.

Blood Pressure	Your blood pressure over time tells us a great deal about your health and relate directly to the proper operation of your heart and lungs.

Heart Rate	Tells us about your overall cardiac and lung condition. A normal heart rate is 60-100 Beats per minute.
Lung (Respiratory) Sounds	If you cup your hands over your ears and breathe in deeply (inhale) the sound you hear is what we listen for within your lungs (breathing through your nose works the best). You tried it, didn't you? It is true, it works. Any other sound is considered abnormal. Some noises are faint and some are quite loud. A normal breathing (respiratory) rate is 12-20 breathes per minute.
Physical touch (palpitation)	Cold or warm hands and feet, how your skin and veins look, the color or your lips and fingernails (not the polish) and/or any vibration felt over the chest.

Testing Procedures

There are many tests and testing procedures, everything from Arterial Blood Gas (ABG), Electrocardiogram (EKG) all the way through the alphabet to Ventilation Perfusion Scans (V/Q Scans). Many of these tests and procedures relate to specific conditions, diseases or disorders so, we will discuss them later in the chapter.

Blood testing is a great tool used to diagnose or rule out diseases or infections.

Note: A new blood test to detect cancer cell may be available soon. It will allow your doctor to treat the disease early and indicate how well the treatment is working which will allow the proper changes in treatment.

Patient Assessment (a more in-depth look)

Most people reading this book now have a basic understanding of the patient assessment and can move on to chapter 9. For those who need a more in-depth understanding or those who are researching something specific will want to read the remainder of this chapter. The information from here to the end of the chapter is in much more depth but is still at a basic level. It has been included to support patients that require additional information. You may want to scan over this information because although it is basic information it can become challenging.

Bedside Assessment, Observations and Interview

Upon entering a room, your healthcare professional will begin to make observations and obtain his or her first impressions. Observing one's respiratory rate, chest and neck movements, oxygen use, speech, cough, level of consciousness (how alert you are or sensorium), items on the bedside table, and most importantly, things that are not on the bedside table (i.e. incentive spirometer, inhalers, etc) are all clues that must be gathered in the initial assessment of a patient. A discussion of your health history and any amplifying factors (allergies, sleep, exercise, etc.) are instrumental in determining the fine points of treatment. The most important factor in the interview is your frank communication of information. You must be an active participant: no one knows better, how you feel and what works or does not work for you. As you speak, we will observe you for any change in speech, changes in breathing rate and how many words you speak between breaths (dead give away to breathing problems or if you play cards, its your tell).

Your physical appearance holds important clues to your health. An inspection for edema (retained fluids), dry mucous membranes (dry mouth, nose or eyes), skin turgor (gently pinch the back of your hand, your skin should return to its normal position and not stay pinched up which can identify dehydration), facial expressions (i.e. pain, fear or work of breathing), warmth and color of your skin. Visually we check for proper oxygenation by looking for cyanosis (a blue tint to skin, lips or fingernails), and capillary refill (gently pinch your fingernail and the pink color should return quickly); these are indications of perfusion (blood supply to the small capillaries). Additionally, a shifting of the trachea (windpipe out of place) could indicate serious lung problems (see chapter 9). The way you sit (tripod or posturing) or the use of accessory muscles (chest wall, neck, abdominal etc) to assist your breathing indicates how hard you are working to breathe or work of breathing (WOB) and help determine if our healthcare plan is working. It is important to note that physical signs and symptoms can change rapidly.

Note: Identifying dehydration early prevents the need for IV fluids.

Clubbing (fingers and toes)

Clubbing is a distortion (bulbous appearance or fattening) of the ends of the fingers and toes. It also causes long smooth curved fingernails and toenails. It is commonly found in patients with lung disease where blood oxygen levels are lower than normal over a very long period of time. It is important to note that clubbing is caused by many causes (etiologies).

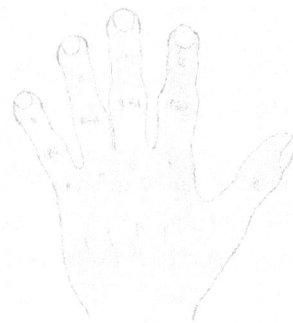

Clubbing

Blood Pressure

Everyone has had his or her blood pressure checked. Blood pressure (BP) is the pressure exerted on the walls of blood vessels by blood circulating during and after it has been pumped by the heart. Systolic pressure (the top or larger number) and is measured while the heart is beating. The diastolic pressure (the bottom or lower number) and is measured while the heart is at rest. It is routinely measured at the brachial artery (your upper arm) using a sphygmomanometer (a pressure cuff and gauge) and stethoscope. There are many ways to measure blood pressures requiring an IV line or by simply palpating (feeling your pulse). The palpation method gives a trained professional a quick estimate of blood pressure (i.e. at what pressure your pulse is lost; Radial pulse (wrist) 80mmHg, femoral pulse (inner thigh) 70mmHg carotid pulse (side of neck) 60mmHg). These estimates are usually higher than your actual blood pressure (but it is great for a quick assessment). Your body and each of its organs have specific blood pressures that we will not discuss here.

Everything influences your blood pressure. If it is inside your body, outside your body, any movement and any thing you see smell, hear, eat or think about, even sleep apnea changes your blood pressure. That white lab coat coming over will also alter your normal blood pressure. Even though your blood pressure may be outside the normal values, your healthcare provider may consider it normal for you.

Classification of Blood Pressure		
Category	Systolic	Diastolic
	mmHg	mmHg
	Top #	Bottom #
Hypotension	<90	<60
Normal	90-119	60-79
Pre-Hypertension	120-139	80-89
Stage 1 Hypertension	140-159	90-99
Stage 2 Hypertension	$\geq$160	$\geq$100

Heart Rate

Heart rate is how fast your heartbeats per minute (60 to 100) is considered normal. A heart rate of <60 is bradycardia (slow) and >100 is tachycardia (Fast). It is routinely measured by counting your heart beats for 15 seconds to 1 minute (1 minute is much more accurate). Everything influences your heart rate. If it is inside your body, outside your body, any movement and any thing you see smell, hear, eat or think about, even sleep apnea changes your heart rate. That white lab coat coming over will also alter your normal heart rate. Even though your heart rate may be outside the normal values, your healthcare provider may consider it normal for you.

Everyone medical professional has a stethoscope, but what are we trying to hear? Listening or auscultations for normal and abnormal heart and breath sounds are very important in determining what is wrong and how to treat your condition. Abnormal breath sounds and causes for each are listed below:

Respiratory (Lung) Sounds

Below you will find a list most of the abnormal breath sounds during respiration. Many can be heard without a stethoscope however, the stethoscope allows us to locate the origin of the sounds within your lungs and chest.

Auscultation   **Sounds and Cause**

Breath Sounds   Sounds are checked at least over each lobe if not
                                                       each lung segment (see chapter 7). If you put

	your fingers in your ears and breathe deeply that is similar to the sounds we listen to hear.
Rales & Crackles are the same thing	Wet lungs (retained fluids), excessive secretions (mucus) which usually clear after a cough or can be caused by collapsed airways.
Fine Crackles	Sudden opening or the peripheral (small airways deep in the lungs) and can be caused by Atelectasis (airless portion of the lungs), fibrosis (unusual fiber tissue growth), and/or edema (fluid retention). See chapter 9.
Coarse Crackles	Movement of secretions within the larger airways, caused by pneumonia or bronchitis (see chapter 9).
Wheeze	Narrow airways due to constriction, swelling, mucus or foreign body (something that does not belong in the lungs, like food). The higher the pitch the smaller the airways have become.
Inspiratory Wheeze (breathing in)	Narrow airways.
Expiratory Wheeze (breathing out)	Narrow airways usually associated with asthma or bronchitis.
Rhonchi	Due to thick secretions, muscle spasm, and/or external pressure (outside the lungs). Sounds are low pitched and continuous. Used to be referred to as low-pitched wheezes.
Rhonchi Fremitus	Similar to tactile Fremitus (vibrations felt on the chest), caused by air passing through narrowed airways.
Diminished	Reduced airflow to lungs due to a restriction or obstruction inside or outside the lungs.
Stridor	Obstruction of the upper airway usually caused by Croup, Epiglottitis (see chapter 9) or after extubation (removal from a ventilator).
Tactile Fremitus	Patient counts softly 1-2-3 while chest is palpated (felt for vibrations). If heavy vibrations are felt lung consolidation (more solid tissue due to fluids or tumors etc) is most likely present. If nothing is felt, it could indicate a pneumothorax (air between the chest cavity and lungs).
Pleural Friction Rub	Sounds like a boat's line rubbing together or creaking sound like two balloons rubbing together that does not go away when you cough. This condition is known as Pleurisy (lack of lubricating fluids between the lungs and the chest wall). See chapters 9 & 11 for details.
Bronchophony	An increased transmission of your voice through the lung tissue when tissue consolidation is present.
Egophony	The patient whispers the letter "e" and due to dense tissue (consolidation), the frequency

	transmitted to the chest sounds like the letter "a".
Whispering Pectoriloquy	1 2 3 is high pitched with consolidation due to more dense tissue and decreased resonance over a pneumothorax (air trapped between the lungs and chest wall, see chapters 9 & 11).
Percussion of the chest	Tapping on the chest creates a low pitched and dull sound (normal resonance) with normal tissue (see chapters 9 & 11).
Tracheal Sounds	Upper airway edema (excess fluid), restriction or obstruction.

Chest excursion (chest movement) and respiratory patterns or breathing patterns are import also (see figure below).

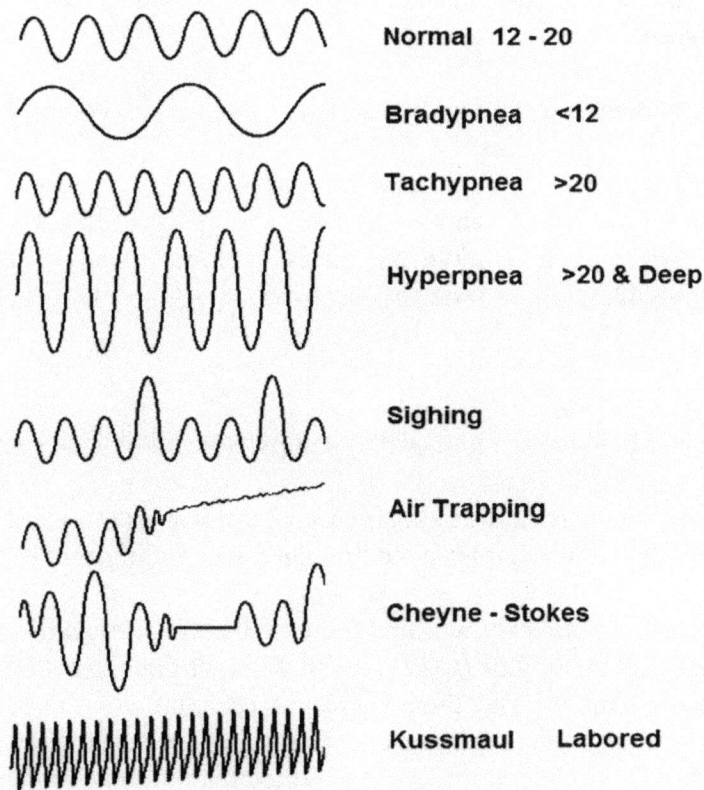

Respiratory Patterns

Heart Sounds or Heartbeat

Anyone who has ever held a stethoscope has listened to their heart to hear that lub-dub sound repeating over and over. There are many heart conditions, diseases and defects which make specific noses and that is why your healthcare provider listens to your heart.

Note: The same is true with bowel sounds.

This is all done within just a few moments of meeting the patient. What is next?

Arterial Blood Gas (ABG)

The Arterial Blood Gas (ABG) checks how well the lungs are functioning. The human body likes things in a status quo (normal), medical professionals refer to it as homeostasis. The body will adjust heart rate, respiratory rate, blood pressure and other functions to ensure enough oxygen is supplied to each cell of each organ for any event that is happening. It will also increase breathing rate to lower the acidity level or slow the breathing rate to increase the acidity of the body as necessary. There are many more checks and balances within the body but we will limit the ones discussed for the most part to basic pulmonary and cardiac related items for our discussion.

Note: Everyone will tell you that an ABG has to hurt a lot, but this is not necessarily true. It really depends on two things; the skill of the healthcare professional drawing the blood and the luck of not hitting a nerve. We will discuss the procedure in detail later in this chapter.

Assessing the ABG can be very complicated, so we will review the basics. Let us discuss normal values first (see table below):

Label	Description	Normal ranges
pH	acid base balance	7.35 – 7.45
PCO2	carbon dioxide	35 – 45
PO2	oxygen pressure in blood*	80 – 100mmHg
SaO2	oxygen saturation (sat)	95 – 100%
HCO3	bicarbonate in blood	22-26 mEq/L
BE	base excess	$\pm$2 mEq/L

*Oxygen based on being at sea level without supplemental oxygen (extra oxygen).

When the pH is higher than normal, it is alkaline in nature, like increasing the pH in a pool. When the pH is lower than normal, it is acidic in nature, like adding acid to a pool.

If the pH is too low (acidic), the body will first increase its rate of breathing. This will cause your lungs to blow off carbon dioxide (CO_2), which will increase the pH by reducing the acids (basically carbon). Another way the body can increase the pH is to add bicarbonate of soda. Just like, you would add soda ash to a pool to raise the pH. If the pH is too high (alkaline), the reverse will take place slowing down the breathing rate to increase the acid level within the body or reducing the bicarbonate of soda. These are classic respiratory related acidosis or alkalosis.

There is another type of pH imbalance, metabolic acidosis/alkalosis and this can be caused by kidney malfunction (a vital organ in controlling pH balance) diabetes or many other problems. It is much harder for the body to correct and most always requires medical intervention. Additionally, just to complicate things, metabolic and respiratory acidosis/alkalosis can be mixed together.

Note: ABG's (Arterial Blood Gases) do not feel good, but they are not supposed to hurt much more than a shot. The ABG procedure is will be discussed in chapter 11.

Note: Special assessment testing will be covered in chapter 11.

Notes:

Chapter 9
Diseases & Treatments

In this chapter we will begin with some basic terms and facts to help you understand diseases and treatments.

It is important to understand that a bacterial infection and a viral infection may seem similar due to symptoms however; they are not the same thing.

> **Bacteria's or bacterium** are microorganisms with a wide range of shapes (i.e. spheres, rods and spirals) and are 100 times larger than a virus. There are approximately five nonillion (that is 5 x 10 to the power of 30) bacteria on our planet. Many types of bacteria benefit the body and its functions. The study of bacteria or bacteriology is a branch of microbiology. Antibiotics are used to treat bacterial infections and antiviral medications have no effect on them.

> **Viruses** are microscopic infections (Type A & B) which can only reproduce inside a host cell. There are more than 5000 types of virus on earth varying in shapes (helical & icosahedral to more complex structures) and 1/100th the size of bacteria. Viruses use their DNA or RNA genes to infect or change a host cell and most can not be eliminated by your immune system. The study of viruses is known as virology and is also a branch of microbiology. Antiviral medications are used to treat them and antibiotics have no effect on them. Every year the flu season overloads our hospitals (approximately 2 million patients annually) typically resulting in 30 to 40 thousand deaths nationwide. If your healthcare provider recommends a flu vaccine, take the immunization.

Viruses (Just the basics)

A virus is a small called a microorganism (can only be seen using a microscope) infectious agent that can only replicate inside the cells of the host organism.

There are three types of influenza viruses; types A, B and C. Types A & B are the human influenza viruses that cause severe seasonal flu epidemics and Type C infections only cause a mild respiratory illness.

Type A viruses have a protein coating (a capsid) which are called Hemagglutinin and Neuraminidase (surface proteins). The H for Hemagglutinin and N for the Neuraminidase make up the nomenclature of the virus (H1N1 Swine Flu, H5N1 Bird Flu etc.). There are 16 H and 9 N combinations however; few are commonly found in humans. The Influenza A viruses are divided into about 25 subtypes and further broken down into different strains. Influenza B viruses are not broken down into subtypes however they are broken down into different strains. Influenza C viruses are not included into the annual flu immunizations due to the mildness of the illness.

Pregnancy suppresses the immune system (the body has a hard time fighting bacteria's and viruses) and restricts the movement of the lungs (because the baby is in the way). This means you are more susceptible to getting sick for two reasons suppressed immune system (that's obvious) and you can not cough as deeply to expel mucus from the lungs.

The common cold is brought to us by over 200 types of viruses; Rhinovirus which cause about 50% of colds, Corona viruses (named for their look, corona of the sun) which cause about 25% of colds, and Respiratory Syncytial Virus (RSV) which cause about 10% of colds.

Note: Flu season occurs during the cold half of the year based on the hemisphere (above or below the equator) in which you live. Flu season begins in November and ends in March in the United States. They are not due to the old wives tale, wet hair, wet cloths or wet shoes.

Vaccinations

Vaccines are biological preparations that improve your immunity to a particular disease. A vaccine typically contains a small amount of an agent that resembles a microorganism. The agent stimulates the body's immune system to recognize the agent as foreign, destroy it, and "remember" it, so that the immune system can more easily recognize and destroy any of these microorganisms that it later encounters. As viruses change (mutate), the vaccines must also be updated for the annual immunizations.

Vaccines come in many different types and based on the strategies designed to reduce the risk of illness and stimulate your body's immune response to the virus. Your immune system Is designed to recognize the virus or vaccine you are exposed to, destroy the virus, remember and recognize the virus the next time you are exposed.

Vaccines do not guarantee protection from a disease. If you have immune system problems or the virus mutates (changes) you can get sick. Vaccines or immunizations stimulate our immune system giving our bodies the ability to fight off infections. It is generally thought that the more flu shots you receive during life will give you protection against respiratory ailments.

Vaccines or immunizations stimulate our immune system giving our bodies the ability to fight off infections. It is generally thought that the more flu shots you receive during life will give you protection against respiratory ailments. Scientists are currently working on a vaccine that will target parts of the virus that do not change (mutate) routinely.

It is important to understand that all vaccines are different. Some last only a year and others can last a lifetime. Talk to your healthcare provider about your needs.

Vaccine production is done through several stages using chicken eggs or grown human cells. In addition, Molecular Flu Vaccine Research or Molecular Virology, the study of viruses at the molecular level (two or more atoms combined or molecules) studies the individual viral genes of a virus and its gene products (after interacting with your body). Researchers are working toward faster and cheaper ways of making flu vaccines.

Note: Acetaminophen or paracetamol (like Tylenol) or other cold/muscle/flu medication usage may interfere with your immune system and should not be taken shortly before or after a vaccination.

The Food and Drug Administration (FDA) has approved a new vaccine that protects against more strains of bacterial infections (about 13 in total). A reduction about 75%) in the number of cases of bacterial pneumonia and ear infections should be eliminated in children. The

CDC (Center for Disease Control) recommends that everyone receive the vaccination (the "Universal Flu Vaccination"). If everyone received the vaccination far less people would die during flu season. You need to take vaccinations very seriously and talk to your healthcare provider about your vaccination schedule. The new H1N3 swine flu vaccine is now combined with annual flu A & B vaccines, ask your healthcare provider what he/she recommends for you and your family.

Note: Even your dog can experience the flu (influenza "A" virus, H3N8). They have no immunity to the virus however, there is a vaccine available.

Note: Researchers have developed vaccines for specific types of cancer and they are working on preventative cancer vaccines (vaccines that would prevent you from getting cancer).

Titer

A blood test called a titer test is used to see if you have antibodies that fight a specific virus. If your doctor has told you have had a specific flu (i.e. Swine, Bird, Asian, Spanish etc.) ask how they can be sure. You can request a titer test for a specific flu; however, your insurance may not pay for the test (about $300.00+) and it may be cheaper to take the immunization for protection.

Preservatives (in medications)

Many vaccines require preservatives to prevent serous adverse effects (contamination like staphylococcus) forming in the vials. Several preservatives are available (thiomersal, phenoxyethanol, formaldehyde etc.) to lengthen shelf life, improve vaccine stability and protect against bacterial infections. Thiomersal is the most effective; however, it contains some mercury that inaccurately linked to autism. Although there has been no convincing scientific evidence supporting the claim, many manufacturing companies no longer use thiomersal as a preservative in vaccines. If you have concerns, you should discuss this with your healthcare provider.

Categories of Disease (Disease Categories)

There are three broad categories of disease states.

"Acute" which indicates a rapid onset and usually of short course.

"Subacute" of longer duration or less rapid change.

"Chronic" which indicates indefinite duration or virtually no change. Chronic refers to something that continues or persists over an extended period, maybe over a lifetime and may never be reversed.

Pandemic is an epidemic of infectious disease that is spreading through human population across a large region (continent or worldwide). Below you will find the phases of an influenza pandemic:

Pandemic Influenza Phases	
Phase	**Description**
1	No animal influenza virus circulating amount animals have been reported to cause infection in humans.
2	An animal influenza virus is circulating in domesticated or wild animals and is know to have caused infection in humans and is therefore considered a specific potential pandemic threat.
3	An animal or human-animal influenza virus has caused sporadic cases or small clusters of disease in people, but has not resulted in human-to-human transmission sufficient to sustain community-level outbreaks.
4	Human to human transmission of an animal or human-animal influenza virus able to sustain community-level outbreaks has been verified. Significant increased risk for pandemic although it has not begun.
5	Human-to-human spread of the virus in two or more countries in one world region. Strong indication that a pandemic is imminent.
6	In addition to Phase 5 criteria, the same virus spreads from human-to-human in at least one other world region.
Post Peak Period	Levels of pandemic influenza in most countries with adequate surveillance have dropped below peak levels.
Post Pandemic	Levels of influenza activity have returned to the levels seen for seasonal influenza in most areas.

Note: There have been many pandemics throughout the world (i.e. Spanish flu, Asian Flu, Hong Kong Flu, Bird Flu, Swine Flu, Seasonal Flu, etc.). Some were really bad and some were not so bad.

Stem Cell Research

There are two types of stem cells; adult and embryonic. Both show great promise as building blocks for life through their ability to transform into 220 different tissues within your body. The goal is to use your own cells to replace worn out organs or vessels which have been damaged by disease or trauma.

Stem cell treatments are at this time limited to a few blood diseases however; in the future, they may be the key to curing many diseases. Please do not be taken in by unscrupulous scam artists and swindlers who are only interested in your money. Discuss any such treatments with your healthcare provider and research it yourself with the FDA (Food and Drug Administration) before wasting your time and money or putting your life at risk.

Note: Laws prohibit creation of embryonic stem cells for research.

Diseases and Treatments

We have created an alphabetical listing with definitions, causes, symptoms, diagnostic tests and treatments for many adult, pediatric (childhood) and neonatal (newborn) diseases. We have provided this information in relatively simple terms, in hopes that everyone will be able to understand and begin the mental preparation of their disease management. If additional questions should arise, please consult other sources for clarification, particularly your Doctor, Nurse and Respiratory Therapist. Additionally, one will also find a tremendous amount of detailed information on line (http://www.webmd.com, http://medlineplus.gov/, http://www.nih.gov/ etc,) or on many local hospital websites although you may find the terminology a bit difficult.

If you wish to accumulate other sources of reference material, we recommend the Merck Manual for descriptions and treatment and the Taber's Medical Dictionary for the definitions of terms that may not be familiar. Medicine changes routinely and the medical reference materials you collect are expensive and will require replacement in order to keep information current.

It is important to remember; the intention of this book is to help you understand your condition and help you decide what questions to ask your healthcare providers. One should have expectations of improved health, and quality of life. This can be accomplished through diligent study, by asking informed questions and by proactively working with one's healthcare providers. This information is not designed to assist you in self-diagnosis or to help second-guess your healthcare professional. If you have any doubts concerning your diagnosis or treatment, getting a second opinion has never been a bad idea.

Abdominal Wall Abnormalities - NEONATAL

DEFINITION   A birth defect (congenital abnormality).
CAUSE   Tumor or swelling (omphalocele).
SYMPTOMS   Severe breathing difficulty (dyspnea).
TESTS   Chest X-ray or CT scan to locate the area.
TREATMENT   Mechanical ventilation.

Acquired Immune Deficiency Syndrome (AIDS)

DEFINITION   An infection by one of four Human Immunodeficiency Virus (HIV or simian & HIV-1 or gorilla, see HIV), which suppresses the immune system and leads to sickness due to opportunistic infections (could be weak germs that attack while the body is in a fragile state).

CAUSE   Presently four different strains of HIV (three types) and HIV-1 (one type) retroviruses which cause AIDS. The virus acts as an immune system suppressant by infecting T-helper lymphocytes (white cells that attack germs) and other cells. HIV is spread through contact with blood and bodily fluids. Unprotected sex, sharing needles, blood transfusions (have all but been eliminated), occupational exposure (working with exposed patients or the virus), and to the baby during birth are all methods of exposure.

SYMPTOMS   Symptoms are similar to having a cold or the flu. Severe weakness, fever, swollen glands, persistent diarrhea or bloody stools, unexplained bleeding, prolonged loss of appetite, weight loss, abnormal decease in white blood cells (leukopenia), and frequent sickness or infections.

Symptoms can take years to develop and in most patients AIDS develops within 10 years of exposure to HIV.

TESTS   Laboratory testing for antibody: ELISA and Western Blot (antibodies develop within 6 months) and the use of blood viral load testing (test for the progression of the specific virus).

TREATMENT   Prevention is the best bet. Your doctor will order medications & treatment (prophylaxis) of the primary infection (cause of getting sick with the virus) and secondary infections (caused by being sick). Clearing of mucus and particles from the lungs (pulmonary hygiene), nebulizers and MDI's (bronchodilators), and supplemental oxygen will be ordered as required. Antiretroviral Therapy (ART) is capable of arresting the HIV progression, decrease mortality and has proven itself to have outstanding preventive effects. Additionally, antibiotics and HIV/AIDS specific medications are of great value. Cancers may be treated with chemotherapy, radiation, surgical resection, lung lobe removal (lobectomy), and/or lung removal (pneumonectomy) these surgical procedures are detailed in chapter 11. New medications are being developed as vaccines (2-4 doses may be required) for specific strains of HIV may help to decrease HIV infection risks.

Note: New vaccine testing for HIV is showing great promise in preventing infection.

Acute Chest Syndrome

DEFINITION   A vaso-occlusive (blockage of veins) crisis of the pulmonary (lungs) vasculature (veins).

CAUSE   Commonly seen in sickle cell anemia patients and caused by lung infections due to bacterial, viral, and/or fungal infections.

SYMPTOMS   Fever, productive cough (sputum production), dyspnea (difficulty breathing or shortness of breath) and hypoxia (reduced oxygen level in the blood). It is a common cause of death in sickle cell patients relating to pulmonary infarction, fat embolism and pulmonary edema.

TESTS   Your healthcare provider may order laboratory blood cultures, sputum cultures, bronchoscopy (see procedures) in order to visualize lung tissue and chest a X-ray or CT Scan to detect new infiltrates (white patches)

TREATMENT   Your healthcare provider will prescribe antibiotics, antiviral and/or antifungal medications as required. Fluid management, supplemental oxygen, bronchodilators, Spirometry and chest physiotherapy are essential to managing your quality of breathing. There is some evidence blood transfusions may be beneficial in improving symptoms.

Adenoma (see Bronchogenic Carcinoma)

Adult Respiratory Distress Syndrome (ARDS) also in NEONATE

DEFINITION   A respiratory failure with life threatening respiratory distress usually associated with pulmonary injury. It is a restrictive pulmonary disease due to edema (fluid in the lung tissue). It is also called noncardiogenic pulmonary edema (fluid build up in the lungs not related to the heart), shock lung, wet lung, or pump lung.

CAUSE	Lung tissue damage associated with sepsis (infection spreading into the blood), viremia (virus in the blood), thoracic trauma(chest damage), hemorrhagic shock (loss of a lot of blood), pancreatitis (inflammation of the pancreas), air or fat emboli (fat or air in the blood in the lungs), aspiration pneumonia (sucking mouth or stomach contents into the lungs), near-drowning (water inside the lungs), oxygen toxicity (too much oxygen), or prolonged mechanical ventilation resulting in inadequate surfactant (lung lubricant) activity.
SYMPTOMS	Difficulty breathing (dyspnea), rapid breathing (tachypnea >20), rapid heart rate (tachycardia >100), cyanosis (a blue tint to skin, lips or fingernails), and hypoxemia (low blood oxygen level). Breath sounds include rales (wet sounding) and wheezes. Severe hypoxemia (very low blood oxygen level), pulmonary edema (fluids within the lung tissue), pulmonary fibrosis (abnormal fiber tissue), and/or respiratory failure.
TESTS	Pulmonary Function Tests values are below predicted values and ABG testing for hypoxemia (low oxygen level in the blood), or hypercapnia (high carbon dioxide or CO2 level). Various laboratory blood tests. X-ray shows diffuse pulmonary infiltrates (junky looking lungs) with ground glass or honeycomb appearance (atelectasis) caused by collapsed or airless areas within the lungs.
TREATMENT	Treat underlying causes and complications. Mechanical ventilation as required to treat elevated carbon dioxide levels causing respiratory acidosis (increased acidity of the blood). Oxygen is administered as the lowest amount necessary to maintain an acceptable oxygen saturation (checked by the finger probe that glows red). Steroids (to reduce swelling), antibiotics (for infections), diuretics (help remove excess fluids), surfactant therapy (to replace lung lubricant), and/or Extracorporeal Membrane Oxygenator (ECMO), will be ordered as indicated.

Note: ECMO is a machine that removes blood from the body, adds oxygen to the blood and then puts it back into your body.

AIDS (see Acquired Immune Deficiency Syndrome)

Airway Obstructions NEONATAL

DEFINITION	Congenital Abnormalities (birth defects) with extra or missing connections between the esophagus and trachea.
CAUSE	External obstructions (outside the lungs) and internal obstructions (there are five types of Tracheoesophageal Fistulas) limit the ability to move air. Fistulas can be complicated and should be researched separately as required.

Note: Normally you swallow using your esophagus and you breathe through your windpipe (airway or trachea). When you have a fistula you have an unwanted pathway or pathways between windpipe and swallowing tube. Additionally you could have a gap in your swallowing tube (esophagus) which may or may not attach to your windpipe (airway or trachea). If food or gastric juices' get into your lungs you will develop what is called Aspiration Pneumonia (see chapter 9). See the drawing below:

Trachea
Esophagus
NORMAL

Fistula
Only

Atresia
With
Lower
Fistula

Atresia
Only

Atresia
With
Upper
Fistula

Atresia
With
Double
Fistula

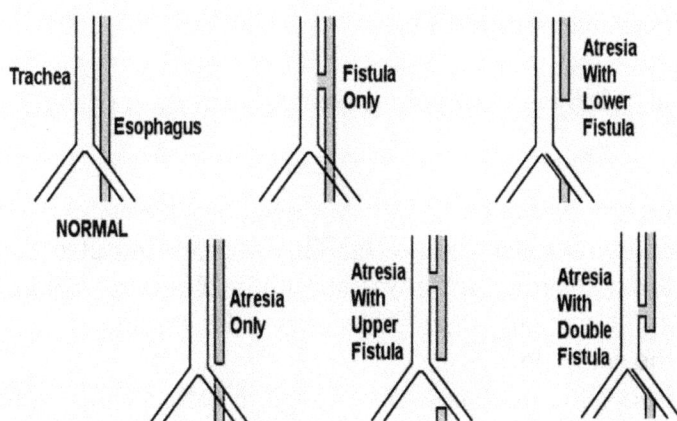

Note: Locating drawings of these abnormalities can be difficult to find, so we drew them for you. The gray is the food tube (esophagus) and the white is your windpipe (airway).

SYMPTOMS   Difficult breathing (dyspnea).
TREATMENT   Supportive care (i.e. oxygen, high frequency ventilation etc.) and/or surgery.

Allergy Management in Asthma (Allergic Asthma)

Allergic reactions are the immune system's reaction to an exposure that would normally not cause you any problems. The management of asthma in the long-term requires the identification of the allergens and irritants which trigger the asthma attack. The Allergen skin tests; skin prick test or intradermal (injection) test play a large role in identifying what allergens adversely affect you. Allergy shots have been proven to reduce the frequency of reactions or attacks and reduce the costs of care (see Asthma).

Alpha Antitrypsin Deficiency (see Emphysema)

Alveolar Proteinosis

DEFINITION   A progressive and chronic deposition of lipoprotein (fat proteins) into alveoli (in the small air sac tissue of the lungs).
CAUSE   Idiopathic (no clear causes).
SYMPTOMS   Can be asymptomatic (no symptoms) or can become fatal. Shortness of breath (dyspnea), cough, digital clubbing (fingertips are fat and nails are rounded over time), and low blood oxygen level (hypoxemia) are the most common symptoms.
TESTS   X-ray: may show diffuse bilateral (both sides) infiltrates (junky nondescript white areas) within the lungs.
TREATMENT   Bronchial hygiene (clearance of mucus) with Chest Physical Therapy (CPT) or vibratory vest therapy (vest is powered by air and looks like a life preserver), and producing, through practice an effective cough. Equipment is available to assist you in producing an effective cough (see chapter 6). A Bronchoscopy (placing a small scope into the lungs through the mouth) and with alveolar lavage (washing the lungs) will help in clearing secretions and easing the work of breathing.

NOTE: Bronchoscopy is not a solution, often this procedure will increase to production of secretions over time (see chapter 8 and 11).

Amyotrophic Lateral Sclerosis (ALS) (Lou Gehrig's Disease) (Progressive Bulbar Palsy)

DEFINITION
: Progressive descending (starting at head or neck) degenerative neuromuscular disease (inability to move muscles which gets worse over time) resulting in the wasting of muscles (atrophy) generally after age 40. Eventually the process causes the wasting of the muscles throughout the entire body.

CAUSE
: Idiopathic (not known), although the effects are caused by the progressive degeneration of corticospinal tracts (from the cerebral cortex) and anterior horn cells (motor neurons for skeletal muscles) or bulbar efferent neurons (motor nerves) preventing the movement of the muscles.

SYMPTOMS
: Muscle wasting (atrophy), muscle weakness, cramps, shallow breathing (hypoventilation), ineffective or lack of cough and progression to respiratory failure.

TESTS
: Due to a decreased ability to uses your respiratory muscles a Pulmonary Function Test (PFT) will indicate a restrictive disease with decreased FVC (Forced Vital Capacity). Peak flow and spirometry abilities are decreased (see chapter 6 & 11).

TREATMENT
: Although there is no specific treatment yet; great strides in understanding the process and possibilities of cure may come from the use of stem cell research (not necessarily embryonic cells). Prevention of complications and provide ventilator support as required. Administer oxygen, assist with the clearance of secretions (CPT or Chest Physical Therapy) & suctioning, and assist with cough effectiveness (see chapter 6 for equipment used). Prognosis is poor; most patients will die within 1-5 years due to pulmonary complications.

Note: ALS research indicates an enzyme known as Activated Proteins (APC) may be able to extend patient lifespan by decreasing gene mutations. Currently these medications cause bleeding and sepsis in patients however, the research continues.

Angina (see Coronary Artery Disease for information)

Anthrax (Bacillus Anthracis)

DEFINITION
: A rare acute and lethal bacterial disease (very fast onset).

CAUSE
: Breathing in (inhaling), eating (ingesting) or contact with an open sore. Spores begin to multiply very rapidly and cause the infection. Anything you carry, sit in, or wear can be contaminated by spores.

SYMPTOMS
: Symptoms are dependant on how you were infected. Pulmonary (inhaled) symptoms are flu-like becoming more sever or fatal (respiratory collapse), ingesting (gastrointestinal anthrax) cause throat/mouth sores, severe diarrhea, vomiting of blood, swelling of the intestinal tract, loss of appetite, and skin (cutaneous) anthrax causes

itchy skin, dark blisters, boils with a black center which are normally free of pain.

TESTS Laboratory cultures (growing samples), a gram stain (to identify what it is) of the sample and biopsy of sample are used in diagnosis.

TREATMENT All three can be fatal if not treated with intravenous (IV) and oral antibiotics (i.e. ciprofloxacin, doxycycline, penicillin, erythromycin vancomycin etc.) immediately. The vaccine is the most effective however it must be administered prior to exposure. Isolation is required and mechanical ventilation may e required for sever cases.

Note: Spores can live in very cold (Antarctica) or hot (desert) conditions for centuries.

Apnea of Prematurity NEONATAL

DEFINITION A very common disorder easily controlled and usually resolves over time.

CAUSE Caused by prematurity (early birth or poor growth).

SYMPTOMS Apneic episodes (not breathing) for 5-10 seconds followed by 10-15 seconds rapid breathing (tachypnea is abnormal if >15 seconds). Also abnormal if the apneic periods are associated with cyanosis (a blue tint to skin, lips or fingernails), pallor (paleness), and/or hypotonia (relaxing of the muscles around arteries).

TESTS Observation.

TREATMENT Continuous SpO2 monitoring (finger probe that glows red and checks the oxygen level in the blood), physical stimulation and possibly mechanical ventilation. Management of the specific cause if known. Additionally CPAP (Continuous Positive Airway Pressure), caffeine, doxapram or theophylline via IV or transfusion.

Asbestosis (see Mesothelioma)

Aspergillosis

DEFINITION An infectious disease of the lung caused by Aspergillus fumigatus (fungus) of which there are three distinct types.

CAUSE The opportunist fungus (Aspergillus fumigatus) found in decaying vegetation, grain, and soil, when inhaled or ingested causes lung infections.

SYMPTOMS Include asthma symptoms, productive cough (coughing up sputum), fever, chills, hypotension (low blood pressure), prostration (absolute exhaustion) and dyspnea (shortness of breath or difficulty breathing).

TESTS A chest X-ray will reveal one or more fungus balls. Blood test may reveal an elevated serum IgE levels (immunoglobulin G or concentration allergy tests).

TREATMENT Dependant on type and severity of the infection. Bronchial hygiene therapy (removing secretions), antibiotics (not for the fungus but rather for resulting infections), antifungal medications (to kill the fungus) and/or surgery (lung resection) may be required (see chapter 11).

Asthma

DEFINITION   A major cause or asthma is genetics, but like so many other chronic diseases, it involves a combination of genetics and environmental exposure. It can be mild to severe obstructive pulmonary disease characterized by reversible airway obstruction due to hyper-reactive airways (smooth muscles in the airways constrict) rapidly and the production of mucosal edema (swelling or inflammation) and/or tenacious mucus (very thick mucus).

CAUSE   Asthma (airway constriction and airway inflammation) can be triggered by exercise, dust, smoke, pollution, pollen, animal dander, dust mites, insect parts, tobacco smoke, chemicals, odors, infection, cold, heat, humidity, barometric pressure changes, aspirin, stress, anxiety, some foods, animals, exercise and many other reasons. Intrinsic (Non-allergic) Asthma is not caused by environmental factors, where environmental irritants cause Extrinsic (Allergic) Asthma. Additionally children who are obese are more likely to develop asthma.

Note: The major cause for Asthma becoming a medical emergency requiring an emergency room visit is overlooking the physical or behavioral signs which should alert you.

SYMPTOMS   Cold or allergy symptoms, cough, shortness of breath (dyspnea), rapid breathing (tachypnea >20), chest congestion, and chest tightness or pain. Breath sounds include inspiratory and expiratory wheezes; rales (wet sounding) if pneumonia is present and the sounds of air movement are diminished. In severe episodes, the patient may not be able to speak more than a few words at a time. The increased work of breathing can eventually lead to fatigue, hypoventilation (slow breathing), hypercapnia (increased carbon dioxide levels in the blood), and respiratory failure. Status asthmaticus (acute asthma attack) is severe and can be life threatening and will develop very rapidly.

Note: An Antigen-antibody IgE (immunoglobulin E) reaction occurs causing the rupturing of mast cells. The ruptured mast cells release histamines, kinins, and serotonin. They have a very potent inflammatory effect, which causes increased mucus production, mucus plugs (thick mucus), mucosal edema (retaining water), and bronchospasm.

TESTS   Laboratory tests will indicate an increase of white blood cells (WBC) with infection and Eosinophil (asthma related type of WBC) count elevated in extrinsic asthma. Pulmonary Function Testing (PFT) will indicate an obstructive disease and decreased flow rates due to the constrictions. FVC (Forced Vital Capacity) and peak flow will be significantly decreased (see chapter 6 & 11). The Forced Expiratory Volume (FEV1) is a very good indicator of changes in the breathing ability however it may not correlate with asthma symptom scores. ABG (Arterial Blood Gas) will reveal a higher than normal carbon dioxide level (hypercapnia) and the patient's pH (acid/base balance is checked to determine sufficient severity to warrant hospitalization. Diagnosis involves several components, including an assessment of the patient's medical history, a physical exam, and Spirometry. Exhaled Breathing

Profiling may be useful in detecting volatile organic compounds as biomarkers. These biomarkers may help differentiate between COPD and Asthma through molecular profiling.

TREATMENT   Treatments should be chosen based on an assessment of asthma severity and control. Allergy shots have been proven to reduce the frequency of attacks and reduce the costs of care. Medications (i.e. inhaled corticosteroids, Leukotriene modifiers, non-steroidal anti-inflammatory drugs etc.) are prescribed to suppress inflammation whereas quick relief medications (e.g. short-acting beta-agonists) are prescribed to alleviate bronchoconstriction.

Fast acting bronchodilators (Albuterol, Xopenex etc.) and long acting beta 2 agonists (Spiriva, Atrovent etc.) are used as needed for symptoms in an attempt to reduce the severity and reverse the constrictions. The patient should receive pulmonary education and rehabilitation (how to avoid known allergens/triggers, how to recognize warning signs, prevention, therapeutics, and medications). The symptoms will usually improve after bronchodilator treatments are administered however, steroids are sometimes required. Patients should be monitored closely with questionnaires, Spirometry, exhaled nitric oxide, and other tools. Patient education in a written asthma action plan is a crucial component in asthma control and should include instruction in how to use medication delivery devices.

Note: Keeping the relative humidity of your home and work place at 50% or less will decrease the frequency of asthma attacks and reduce the number of dust mites and others bugs.

Atelectasis

DEFINITION   Acute (comes on quickly) or chronic (longer term) shrunken and airless state of part or the entire lung usually accompanied by infection, bronchiectasis (destructive change in bronchial walls), and/or fibrosis (abnormal tissue growth).

CAUSE   Tenacious mucus plugs (very thick), endobronchial tumors (abnormal growth), or foreign bodies (hot dogs, corn, bread, pudding etc) are common causes. However, pressures from outside the lungs can compress the bronchus and cause atelectasis (airless portion of the lungs). These would include enlarged lymph nodes, tumor, aneurysm (swollen blood vessels that can result in bleeding), pleural effusion (fluid between the lung and chest wall), pneumothorax (air between the lungs and chest wall) and many others.

SYMPTOMS   Dyspnea (shortness of breath or difficulty breathing), tachypnea (rapid breathing >20), chest pain, and tachycardia (rapid heart rate >100). Breath sounds will be diminished over affected area due to the lack of airflow.

TESTS   Chest X-rays reveals white blotching (Radiopacity) or plate like patchy infiltrates with the appearance of ground glass at the affected area. Pulmonary Function testing (PFT) reveal a restrictive pattern (see chapter 8 & 11). If the lung should collapse, a CT scan can help to determine the cause. It may also reveal displaced chest anatomy such as

110

TREATMENT   displaced blood/lymph vessels, possible shifting of lungs and trachea (Mediastinal shift) toward affected side, and/or an elevated diaphragm on affected side.

TREATMENT   Resolve the underlying cause. Incentive spirometer (3-4 time a day), Intermittent Positive Pressure Breathing (IPPB) 3-6 time a day, and/or Positive End-Expiratory Pressure (PEP) therapy. To assist the removal of mucus Chest Physical Therapy (CPT), vibrating therapy vest (powered by air and looks like a life preserver) or Flutter Valve may be prescribed (see chapter 6).

Atherosclerosis or Arterial Disease (also see Cardiovascular Disease or CVD)

DEFINITION: Atherosclerosis (or Arteriosclerotic Vascular Disease (ASVD)) is the condition in which an artery wall thickens as the result of a build-up of fatty materials such as cholesterol, plaques, crystals and/or calcifications (also called atherogenic substances). It is commonly called hardening (loss of elasticity) or furring of the arteries. Also see PAD (Peripheral Artery Disease), PVD (Peripheral Vascular Disease) and PAOD (Peripheral Artery Occlusive Disease.

CAUSE: As described above. There are three terms in describing where this process takes place; (1) Atherosclerosis is hardening (loss of elasticity) of an artery due to plaque build up, (2) Arteriosclerosis is the hardening (loss of elasticity) of medium or large arteries and (3) Arteriolosclerosis in the hardening (loss of elasticity) of small arteries. Things that place you at risk are; high cholesterol, overactive thyroid (hyperthyroidism), diabetes, obesity, smoking, high blood pressure (hypertension), aging, heredity, being male, sedentary lifestyle (not exercising), stress, sleep disorders, and some Sexually Transmitted Diseases (STD).

SYMPTOMS: Typically asymptomatic (having no symptoms) for decades, the slow process is progressive and cumulative. Symptoms vary from artery enlargement, poor circulation (due reduced blood supply or stenosis), pain, swelling of extremities, heaviness in extremities (arms and legs), aneurysm (swollen blood vessels) and/or clots. A clot (thrombus) that breaks free is a medical emergency which can cause heart attack, stroke and pulmonary embolisms. As the rupture or clot heals the scarring will further close the artery due to narrowing (stenosis).

TESTS: Cardiac Stress Testing is the most common non-invasive testing for blood flow limitations. Medical imaging such as angiography (x-ray using a contrast agent or fluoroscopy), ultrasound and/or CT Scanning. Laboratory blood tests may be ordered (D-dimer, HbA1c, hs-CRP, homocysteine levels, lipoproteins etc).

TREATMENT: Prevention is the best bet however; diet, vitamin supplements, medications or surgery (repair and/or angioplasty) may be required to treat the disease and symptoms. Remember it took a lifetime to develop.

Avian Influenza (Bird Flu or H5N1)

DEFINITION
This virus is found in wild birds although they do not usually get sick from the virus.

CAUSE
There are many types of bird flu however, the H5N1 avian influenza virus is the one that concerns us here. It is spread to chickens, ducks and turkeys through bird droppings and saliva. It may also be spread by touching contaminated equipment or eating poorly prepared poultry products. Recent studies have indicated the virus can pass through the placenta to the fetus. Although the virus can grow in the upper airways, it grows best in the lower (smaller) airways in children.

SYMPTOMS
To date, no human infections have been reported in the United States or Canada. Flu like symptoms such as fever, cough sore throat, muscle aches and eye infection (conjunctivitis). More serious cases include pneumonia (lung infection), ARDS (see Acute Respiratory Distress Syndrome which can be deadly.

TESTS
Your healthcare provider may ask where you have recently traveled and if you were in contact with any birds. Blood test, nasal swabs or other tests may help diagnose bird flue. Reverse Transcriptase Polymerase Chain Reaction (RT-PCR) and Real-time Reverse PCR Assay detect viral RNA in specimens or cultures. Additionally, serological identification of MN, HT, EIA, and VN testing are impractical in clinical cases but are very accurate. X-rays will help determine where the lungs have been infected.

TREATMENT
Prevention is the best bet; H5N1 Recombinant Vaccines are available and will protect you from infection. Wash your hands and cook poultry (bird) products well. Antiviral medications may not work against the flu and if used inappropriate may cause other bacteria to become resistant to the medication. If your illness warrants hospitalization you be in isolation to reduce the possibility of spreading the virus. Sever cases may require mechanical ventilation and dialysis (kidney machine to help clean impurities from the blood). Although the FDA (Food and Drug Administration) has approved an immunization, it is not yet recommended for the general public, a vaccination is presently in progress that targets an antibody that can neutralize both strains infecting birds and people. Presently mortality rates are about 50%.

Black Lung (see Coal Workers' Pneumoconiosis (CWP))

Blastomycosis (see Paracoccidioidomycosis for South & Central America)

DEFINITION
An infectious disease primarily involving the lungs but occasionally spreads to the skin.

CAUSE
The inhalation of the fungus (Blastomycosis Dermatitidis) which may be associated with contact with beavers or dogs who carry the infection.

SYMPTOMS
A dry hacking or predictive cough, chest pain, fever chills, drenching sweats, and dyspnea (shortness of breath or difficulty breathing) are the initial symptoms.

TESTS
Diagnosis is certain via a bronchoscopy (small thin scope inserted into the lungs via the mouth, see chapter 6, 8 &11) revealing a thick walled

budding yeast with pus. Serologic tests (testing blood or fluid) are of no value.

TREATMENT Treatment is highly effective and improvement begins within a week. Occasionally, the course of treatment may need to be repeated. Most untreated patients with the disease will slowly progress toward death.

Botulism

DEFINITION A rare and serious condition caused by toxins (poisons) from bacteria (Clostridium Botulinum).

CAUSE Botulism is a high risk for use as a bio-weapon or bioterrorism. Bacteria can be eaten or enter through a wound.

SYMPTOMS Facial weakness, drooping eyelids, blurred vision, difficult swallowing or speaking, difficulty breathing (dyspnea), nausea, vomiting and paralysis (inability to move). Elevated blood pressure (hypertension) and elevated heart rate (tachycardia) are not symptoms and fever is not normally present except wound botulism.

TESTS Laboratory blood tests.

TREATMENT This is a medical emergency and prevention is your best bet (keep food preparation surfaces and utensils clean and was your hands) even during a bioterrorism attack. Antitoxins are available but will not reverse and damage. You may require mechanical ventilation to support your breathing for months. Rehabilitation may be required for speech, swallowing or any function affected by the disease.

Bronchiectasis

DEFINITION An irreversible dilation of the bronchi usually accompanied by infection. It is an obstructive pulmonary disease with destructive changes of the bronchial wall.

CAUSE Congenital (from birth) bronchiectasis is a rare condition in which the lungs fail to develop. Acquired (a new process) bronchiectasis results from destruction of bronchial wall after infection, inhalation of chemicals, immune reactions or vascular abnormalities or secondary to atelectasis (airless portion of the lungs). These result in chronic inflammation, fibrosis (abnormal tissue growth), and impaired mucus clearance (cannot be cleared with a cough).

SYMPTOMS Severity and characteristics vary widely from patient to patient. Chronic productive cough with copious amount of sputum, reoccurring pneumonia, wheezing, dyspnea (shortness of breath or difficulty breathing), right heart failure, and clubbing (fat fingertips with rounded fingernails are typical). Breath sounds reveal wheezes, rhonchi, and rales (see chapter 8).

TESTS The sputum sample separates into three layers; the top layer is frothy and watery, the middle layer is turbid (cloudy), and the bottom layer is opaque (can not see through). The chest x-ray are usually normal however, they may reveal crowded markings ((not enough room for the lungs and tissue), honeycomb appearance, cystic areas (solid due to masses), widespread atelectasis (white patchy plate like, ground glass), and/or cylindrical hazy vascular markings. Over inflation (hyperinflation) of unaffected areas may be present due to air hunger (cannot get enough air). A bronchoscopy (small tube inserted through the mouth into the lungs, see chapter 6, 8 & 11) and a computed tomography (see chapter 11) can reveal location and severity with bronchi clearly outlined. Arterial Blood Gas (ABG) will identify any presence of hypoxemia (low blood oxygen), and hypercapnia (high levels of carbon dioxide) in advanced disease process. Pulmonary Function Test (PFT) will verify an obstructive disease pattern (see chapter 11).

TREATMENT   Control any infections using antibiotics, bronchial hygiene (movement of secretions), aerosol therapy, mucolytics (in order to thin mucus), postural drainage (laying patient in various positions to aid in secretion movement while performing CPT or Chest Physical Therapy), bronchoscopy (see chapter 11) for visualization and suctioning. Surgical resection is rarely necessary except in severe cases where other management efforts are unsuccessful (see chapter 11).

Bronchiolitis Obliterans (see Fixed Obstructive Lung Disease)

Bronchiolitis Obliterans with Organizing Pneumonia (BOOP), (see Pneumonia)

Bronchitis (Acute)

DEFINITION   Rapid onset of inflammation of the bronchial mucosa of the tracheobronchial tree (see chapter 7).

CAUSE   Most prevalent in winter, it normally develops after an acute Upper Respiratory Infection (URI or cold/virus). Additionally, may be caused by dust, smoke, chemical fumes, or ozone.

SYMPTOMS   Cold and flu symptoms; cough, wheezing, fatigue, chills, slight fever, back pain, muscle pain and sore throat. The patient's cough will initially be nonproductive (not coughing up sputum), later becoming productive (coughing up sputum) with increasing in volume and color.

TESTS   Breath sounds reveal rhonchi, and wheezes (see chapter 8). Pulmonary Function Tests (PFT) will reveal an obstructive disease pattern. The Arterial Blood Gas (ABG) may show hypoxemia (low blood oxygen level), hypercapnia (elevated carbon dioxide level), and compensated respiratory acidosis (this is where the body uses bicarbonate or respiratory rate to control its balance of pH, see chapter 11). Chest x-ray is used to rule out other processes and may reveal hyperinflation (over inflation of the lungs) and depressed diaphragm or areas of pneumonia.

TREATMENT   Cough suppressants with expectorants, humidification, oxygen with humidification, bronchial hygiene (moving secretions), and bronchodilators (usually nebulizers) will help with the discomfort. Anti-inflammatory medications, steroids, and antibiotics (for infection) may be ordered.

Bronchitis (Chronic), (also see COPD)

DEFINITION   Long-term inflammation of bronchial mucosa (mucus membrane swollen or wet) of the tracheobronchial tree (see chapter 7). Chronic bronchitis (also called blue bloater) is classified as COPD and will present with productive cough on most days, for 3 months or longer over a two successive years. The airway muscles tighten, airway lining swells and mucus builds up.

Bronchogenic Carcinoma / Lung Cancer

DEFINITION	Lung cancer is malignant tumor or number of tumors resulting from damage to the genetic DNA of cells or mutations in cell lining of the lungs. A tumor (new swelling or enlargement or abnormal growth), neoplasm (a new and abnormal group of cells serving no purpose), or adenoma (tumor on glands) can be cancerous (malignant) or non-cancerous (non-malignant).
CAUSE	Can be from no clear causes (idiopathic), but many cancers linked to carcinogens (cancer causing agents) or other causative substances (ionizing radiation, radon, asbestos, arsenic, some manmade mineral fibers, vinyl chloride etc).
SYMPTOMS	The rapid growth of groups of cells stops or reduces the normal function of the body's tissue. Symptoms will vary greatly due to the type, size and location of the cancer as well as the degree of obstruction. An obstructive tumor can cause dyspnea (shortness of breath or difficulty breathing), cough, hemoptysis (bleeding from the lungs), chest pain, and frequent pulmonary infections due to the obstructions. Breath sounds reveal a diminished sound with wheezes, and/or rhonchi/rales (wet sounding) lungs.
TESTS	Pulmonary Function Tests (PFT) will determine if the tumor is causing obstructive or restrictive airflow but will vary greatly based on the type, severity and location. Chest x-ray may reveal an opaque mass, nodule, infiltrates (junky looking lungs), atelectasis (airless portions of the lung), and if the tumor is large enough a mediastinal shift away from the affected area (the tumor pushes the lungs out of the way).
TREATMENT	Prevention is the best defense (i.e. avoid exposure to carcinogens, ionizing radiation, smoking, etc). Treatment will depend on type, severity and location of the cancer. Any symptoms and/or complications are treated as required with curative or supportive care. Radiation therapy, chemotherapy, surgical resection (reducing lung size), lobectomy (lung lobe removal), or pneumonectomy (lung removal).

Bronchopulmonary Dysplasia (BPD) NEONATAL

DEFINITION	A chronic lung disorder that may occur in infants who require ventilator support due to pulmonary damage or dysfunction (not working correctly).
CAUSE	Associated with low weight preterm infants and barotraumas (damage to lungs) from positive pressure ventilation (mechanical ventilation), high-inspired oxygen concentrations (FiO2), and endotracheal intubation (ET Tube) and is more common among premature infants.
SYMPTOMS	The process may be caused by ventilator related lung collapse (atelectrauma), patient requiring increased levels of oxygen (FiO2) leading to inflammation and lung damage (alveolar damage), and/or

	pulmonary dysfunction. Dyspnea (difficulty breathing or shortness of breath) hypoxemia (low blood oxygen level), hypercapnia (high blood carbon dioxide level) and increased work of breathing (WOB) are common. Additionally, it may begin with pneumonia or sepsis (blood infection) requiring critical care support.
TEST	Chest X-ray may reveal atelectasis (airless areas appearing as white plate like or ground glass or classic pneumonia appearance (white areas)). ABG reveal hypoxemia (low blood oxygen levels) and hypercapnia (elevated carbon dioxide levels) caused by airway obstructions, air trapping, and pulmonary fibrosis (unusual fiber growth in the lungs).
TREATMENT	Preventions is the best bet. Surfactant (lung lubricant) and CPAP (Continuous Positive Airway Pressure) may be sufficient to correct the problem. Recovery may be rapid or require months of ventilator support with the lowest level of inspired oxygen levels (FiO2) as possible. If pneumonia is present, antibiotics will be administered for bacterial infections. Chest Physical Therapy (CPT), oxygen support, bronchodilators (usually via nebulizer) and steroids are normally administered. Some recent studies recommend antenatal steroids be given to women with preterm delivery risks (please discuss risks with your healthcare provider).

Brown Plague (see Dust Pneumonia)

CAD (Coronary Artery Disease)

Cancer / Carcinoma (see Bronchogenic Carcinoma)

Cardiac Arrest (Sudden Cardiac Arrest, Cardiopulmonary Arrest or Circulatory Arrest)

DEFINITION	When the heart stops pumping blood (clinical death).
CAUSE	It is different from a heart attack (thrombosis or myocardial infarction) however, a heart attack can be the cause. It can be caused by heart diseases and non-heart related causes (i.e. trauma (cardiac tamponade or tension pneumothorax), drowning (hypoxia due to lack of oxygen), hypothermia (low body temperature), pulmonary embolism, overdose (medications and/or poisons or neurotoxins), internal bleeding (hypovolemia or low blood level) and/or high or low blood sugar (hyperglycemia or hypoglycemia). Risk factors include heart disease, smoking, obesity, diabetes and family history.
SYMPTOMS	Possible chest pain radiating to the back, left arm and/or jaw, shortness of breath (dyspnea), dizziness and loss of conscientious followed by death if it not reversed.
TESTS	Check for a pulse, listen for the heart beat and monitor heart activity with an EKG.
TREATMENT	Begin CPR (cardiopulmonary resuscitation), assess any heart rhythm with an EKG, attempt defibrillation to restart the heart and administer ACLS (Advanced Cardiac Life Support) medications.

Cardiac Arrhythmias (or Cardiac Dysrhythmis)

DEFINITION   Any of a large group of conditions which cause an abnormal electrical activity of the heart or heart beat (see chapter 8 & 11).

CAUSE   Arrhythmias can be caused by medications, drugs, alcohol, damage to the heart and electrolyte levels (sodium, potassium) which are too high (faster heart rate or tachycardia) or too low (slower heart rate or bradycardia).

SYMPTOMS   There are about 24 different arrhythmias, some being life-threatening and are medical emergencies while some irregular heart beats (palpitations) may just be annoying (however, your healthcare provider should be aware of any palpitations). In serious arrhythmias you may feel the same symptoms as a heart attack (see myocardial infarction). Arrhythmias can be caused by different areas of the heart, atrial (top half of the heart), junctional (electrical disruption between the atrial and ventricular sections of the heart), ventricular (the bottom half of the heart), heart blocks (electrical signal blockages), sudden arrhythmia death syndrome (the heart stops) or any combination of these areas.

TEST   EKG to establish rhythm and damage to the heart, cardiac ultrasound to determine blood flow and laboratory blood tests to detect cardiac enzymes that are released during a heart attack and electrolytes (i.e. sodium, potassium etc.) which cause the heart to beat.

TREATMENT   Treatment will be based on the type and severity of the arrhythmia.

Cardiac Tamponade (see Pericardial Effusion)

Cardiomegaly

DEFINITION   The enlargement of the heart.
CAUSE   Congestive heart failure, Cardiomyopathy and others.

Cardiomyopathy

DEFINITION   Abnormal heart muscle or Primary disease of the heart muscle.

CAUSE   Caused by Coronary Artery Disease (CAD), heart attack, hypertension (high blood pressure), heart valve disease, toxic metals, myocarditis (inflammation of the heart muscles), excessive alcohol consumption, Congestive Heart Failure (CHF), hypertrophy (excessive cell growth) of the interventricular septum (a small wall between the ventricles) or restrictive flow.

SYMPTOMS   Similar to CHF the patient will have chest pain similar to angina, severe dyspnea (shortness or breathe or difficulty breathing), tachycardia (rapid heart rate >100), orthopnea (difficulty breathing when lying down), peripheral edema (swelling in ankles or feet), fatigue (tiredness), dizziness, and syncope (passing out or loss of conscientious).

TESTS   The Electrocardiogram (EKG) and chest x-ray are usually abnormal,

TREATMENT   Treat symptoms of dyspnea (shortness of breath or difficulty breathing) with oxygen or Continuous Positive Airway Pressure (CPAP) see chapter 6, and calcium blockers (like Verapamil). Some patents may require cardioversion (shocking the heart to change the rhythm back to normal) and surgery may be needed in severe cases.

Cardiopathy (see Heart Disease)

Cardiopulmonary Arrest (see Cardiac Arrest)

Cardiovascular Disease (CVD or Atherosclerosis)

DEFINITION   A class of diseases that involve the heart or its blood vessels (veins and arteries). Many types of heart disease exist, and we will discuss the three most common. Heart Attack (MI), Congestive Heart Failure and Congenital Heart Disease.

CAUSE   Very slow onset disease usually progressing for decades and is quite advanced when discovered. The cause will be dependant on the type and severity of the disease (see specific disease). Believed to be caused by unhealthy eating, lack of exercise, smoking and obesity.

SYMPTOMS   A wide range of symptoms are possible dependant on the severity (see note below).

TESTS   Many biomarkers found in blood test may reflect a higher risk High fibrinogen, elevated homocysteine, high C-reactive protein, elevated brain natriuretic peptide and elevated asymmetric dimethylarginine levels.

TREATMENT   Develop a healthy diet, Lowering cholesterol and triglycerides, supplement vitamins B6/B12, increased antioxidant intake, reduce red meats and increase vegetable intake. Unlike many chronic diseases, the disease is treatable and reversible even following a long history.

Note: CVD includes atherosclerosis, stroke, cerebrovascular disease, congestive heart failure, heart failure, coronary artery disease, myocardial infarction and peripheral vascular disease among others.

Chemical/Gas Exposure (Chronic and Acute)

DEFINITION   The long-term or sudden inhalation of chemicals or gases.

CAUSE   The inhalation of sudden release (acute) or long-term exposure (chronic) to gases and chemicals which damage lung tissue.

SYMPTOMS   Symptoms are dependant on which gas, chemical, or combinations of gases/chemicals, length and concentration of exposure, how deeply it was inhaled and any sensitivity you may have to the gas or chemical. Symptoms include cough, coughing up blood (hemoptysis), shortness of breath (dyspnea), pulmonary edema (excess fluids in the lungs tissue), inflammation of the small airways (Bronchiolitis) and/or eye/nose irritation. The long-term inhalation exposure of some chemicals can cause cancer (i.e. arsenic, hydrocarbons, radioactive materials etc.).

TESTS   Chest x-ray, CT scan and Pulmonary Function Test (PFT) in order to determine damage.

TREATMENT   Prevention is your best bet. Treatment consists of supplemental oxygen, bronchodilators (nebulizer), steroids (such as prednisone) to reduce inflammation, antibiotics if needed and mechanical ventilation in severe cases.

CHF (see Congestive Heart Failure)

Chronic Cerebrospinal Venous Insufficiently (CCSVI) see PVD

DEFINITION   A syndrome where the venous system is not able to efficiently remove oxygen poor blood from the Central Nervous System (CBS) due to circulatory valve problems or venous stenosis of the jugular (both sides of the neck) and azygos veins (runs along side your spinal cord).

Chronic Obstructive Pulmonary Disease (COPD)

DEFINITION   COPD is a generalized progressive airway obstruction, in the small airways (lower in the lungs) for the most part. It is associated with combinations of chronic bronchitis, asthma, and emphysema (see chart below). If it is coupled with Obstructive Sleep Apnea it is know as "Overlap Syndrome (see Sleep Apnea).

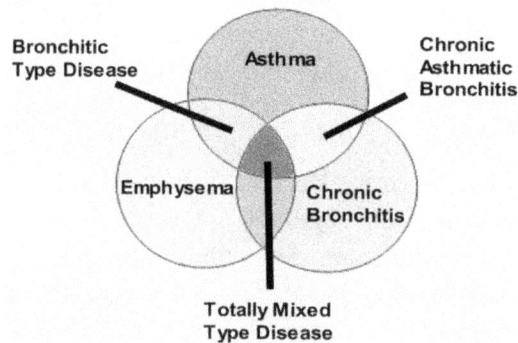

CAUSE   COPD is a culmination of genetics and obstructive diseases taking up to 20 years to develop. The narrowing of the airways caused by the combination of disease processes identified above, and reoccurring exposure to chemicals, pollutants, chemicals, smoke and other irritants. COPD exceeds lung cancer as a major cause of disability and death. Additionally, most other lung diseases are classified as restrictive.

SYMPTOMS   Although COPD is thought to begin early in life, the significant symptoms do not usually occur until middle age and classified as mild moderate or severe. Beginning with mild dyspnea (shortness of breath or difficulty breathing), it will gradually progress to significant dyspnea (shortness of breath) with exercise, to acute respiratory illness. Cough and sputum production are widely varied but are usually significant. Breath sounds will vary from a mild chronic wheeze to extreme wheezing.

Note: Pulmonary Function Testing (PFT) is used to determine COPD severity by measuring predicted values of FEV1 and FVC (see Pulmonary Function Test (PFT).

Mild COPD is ≥ 80% of predicted, Moderate COPD is 50-80% of predicted, Severe COPD is 30-50% of predicted and Very Severe COPD is <30% of predicted.

TESTS	Chest x-ray may reveal a normal chest film early in the disease stage but will progress through over inflation (hyperinflation), depressed diaphragm, generalized radiolucency (brightness or white marking) of the lung fields, increased air space and spreading of the ribs (Barrel Chest). Laboratory test for Serum IgE (immunoglobulin E) and Red Blood Cell are checked for indicators of chronic hypoxemia (long-term low blood oxygen). Pulmonary Function Testing (PFT) will confirm the obstructive disease (see chapter 8 & 11). Exhaled Breathing Profiling may be useful in detecting volatile organic compounds as biomarkers. These biomarkers may help differentiate between COPD and Asthma through molecular profiling.
TREATMENT	There is no cure for COPD, but therapies will relieve the symptoms and will control outbreaks. The sooner you discuss your breathing difficulty with your healthcare provider the sooner your COPD can be managed and you can get back to life. A combination of long acting and short-term bronchodilators as well as the use of steroids can help to relieve the symptoms and reduce the frequency of dyspnea (difficulty breathing or shortness of breath). Patient education and rehabilitation are equally important. Learning to avoid irritants (i.e. smoke, chemicals, etc) is of primary importance. See Smoking Cessation at the end of this chapter.

Note: There is considerable evidence that genetics (inherited traits) may make you more susceptible to COPD processes, especially in newborn COPD.

Note: There is recent research that indicates that COPD is an autoimmune disease, and that it occurs when a person's immune system attacks the cells that line the airways and the air sack of the lungs. The use of inhaled immune-suppressant medications could lead to a better treatment protocol.

Chronic Venous Insufficiency (CVI), (see PVD)

DEFINITION	The veins can not pump enough oxygen poor blood back to the heart.
CAUSE	Deep Vein Thrombosis (DVT) blocking the valves in the circulatory system.

Circulatory Arrest (see Cardiac Arrest)

Clubbing, Digital Clubbing, or Hypertrophic Osteoarthropathy (Acropachy)

DEFINITION	Hypertrophic osteoarthropathy is a disease of the joints and bones. It is characterized by clubbing of the fingers and toes, enlargement of the extremities, and painful and swollen joints. The disease has two types Primary Hypertrophic Osteoarthropathy (an inherited condition) and lungs, heart or liver causes Secondary Hypertrophic Osteoarthropathy.
CAUSE	Secondary Hypertrophic Osteoarthropathy discussed here is associated with diseases, which cause a chronic hypoxemia (long-term low blood oxygen level).

| SYMPTOMS | Chronic inflammation, enlargement and disfigurement of the fingers/toes with obvious lateral (side to side) and longitudinal (length) curvature of the nails and a dramatic increase of soft tissue are quite common. |
| TREATMENT | Treatment of the underlying condition that is causing the problem. Even with treatment of the underlying problem, the clubbing is not reversible. Nonsteroidal Anti-Inflammatory Drugs (NSAID's) may be helpful in relieving pain. |

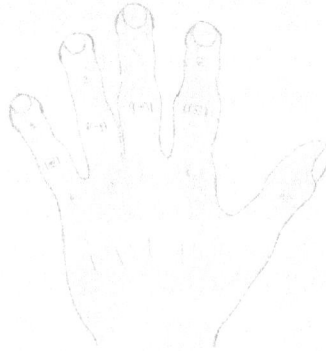

Clubbing

Coal Workers' Pneumoconiosis (CWP)

DEFINITION	Coal dust collects in nodules widely distributed throughout the lungs.
CAUSE	Long-term exposure to coal dust while mining or processing coal.
SYMPTOMS	Because coal is not fibrous, it is not normally associated with fibrodysplasia (abnormal fibrous tissue) within the respiratory system but it does cause stiffening of the lungs (reduced elasticity) making breathing difficult. However, Progressive Massive Fibrosis (PMF) may develop is less than 2% of patients. PMF is a black mass that destroys lung tissue. Symptoms include cough and sputum production.
TESTS	Chest X-ray may show round opacities (white spots) >1 cm (about 1/3 inch) in diameter.
TREATMENT	Prevention is the best bet however the wearing of breathing protection is not practical due to dust clogging the filters and eye pieces. Because there is no treatment available other than the possibility of a lung transplant.

Coccidioidomycosis (Valley Fever or San Joaquin Fever)

DEFINITION	An infectious disease which is a self-limiting respiratory disease which can become fatal even if treated.
CAUSE	A fungus found in soil of Southwest region of North America which when inhaled can cause respiratory distress.
SYMPTOMS	Range from no symptoms (asymptomatic) to serious pneumonia symptoms. Fever, productive cough (sputum), chest pain, chills, sore throat, and/or hemoptysis (blood from the lungs) accompanied by dyspnea (shortness of breath or difficulty breathing) are common.
TESTS	This is an obscure illness. Laboratory culture tests (for fungus) on sputum, gastric washings (from the stomach), pleural fluid (lung fluid) or spinal fluid (CSF) will confirm the diagnosis. Chest x-ray may reveal

thick-walled fungus appearing as white spots, infiltrates (junky looking lungs) or bronchopneumonia.

TREATMENT   Antifungal medication and pulmonary hygiene (moving secretions). Outcome for the primary form is positive however, in the progressive form, the fatality rate is 55-60%.

Cold or Common Cold (see Rhinovirus)

SYMPTOMS   The key symptom is having a runny nose. This symptom alone can eliminate the concerns of H1N1 and H1N3 (Swine Flu) and many others. H1N1 normally will not have the runny nose symptom. All other symptoms see flu.

TREATMENT   Tamiflu will not have any effect on the common cold.

Congenital Diaphragmatic Hernia NEONATAL

DEFINITION   Congenital abnormality occurring one in 2500 live births. The absence of the diaphragm on the left, right or both sided (left sided is the most common) altering lung and organ placement.

CAUSE   Lung hypoplasia (underdevelopment of tissue), decreased alveolar count, a decreased pulmonary vasculature (veins in the lungs), pulmonary hypotension (low blood pressure in the lungs). Additionally, the unusual anatomy of the inferior vena cava, Bochdalek's Hernia (left lateral & posterior), or Morgagni's Hernia (anterior & medial) may be noted.

SYMPTOMS   Severe respiratory distress, scaphoid abdomen (boat shaped), decreased (diminished) breath sounds, displaced organs and severe cyanosis (a blue tint to skin, lips or fingernails).

TREATMENT   Survival is rare. Immediate stabilization and immediate access to specialized ventilation techniques including nitric oxide, ECMO and surgery are critical.

NOTE: ECMO is a machine that removes blood adds oxygen and puts it back into your body.

Congenital Heart Disease NEONATAL

DEFINITION   Defect within the heart at birth.

CAUSE   Presents in two categories, cyanotic (blue tint to skin) and acyanotic (no blue tint) to the skin.

SYMPTOMS   Cyanotic heart disease presents as right to left shunt (venous to arterial leak causing dilution of oxygen in the arterial blood) and acyanotic (absent of blue tint) heart disease presents as a left to right shunt (arterial to venous leak causes not dilution of oxygen however, it does increase the heart's workload significantly).

TREATMENT   Supportive care.

Congestive Heart Failure (CHF or Heart Failure)

DEFINITION   A condition where the heart has lost its ability to efficiently pump blood.

CAUSE	It is caused by many conditions that overwork the heart or cause blood or fluid to back up in the lungs. These conditions include but are not limited to; Hypertension (high blood pressure), Coronary Artery Disease (CAD), Heart Attack (which damages heart muscle due to lack of blood flow), cardromyopathy (infection, alcohol, drug use etc.) and diabetes. Hypertension (high blood pressure), type 2 diabetes, heart failure, stroke, arrhythmias (unusual heartbeat) and nocturnal complex arrhythmias (unusual heartbeat at night) have been associated with Sleep Disordered Breathing (SDB) also know as Sleep Apnea.
SYMPTOMS	Symptoms will vary greatly depending on the severity of the disease process. Symptoms will include; Chest pain or tightness, fatigue (tiredness) due to the body providing blood to the vital organs, dyspnea (shortness of breath) due to edema (excess fluids in the lung tissue), peripheral edema (water retention in feed ankles and legs), persistent productive cough (white or pink frothy sputum), and wheezing. Additionally many patients display orthopnea (difficulty breathing when lying down), weight gain (due to fluid retention), nausea (due to reduced blood flow to the digestive system, confusion, and increased heart rate (due to the heart trying to pump more blood).
TESTS	Laboratory blood tests (to evaluate cholesterol, anemia, kidney and thyroid functions) and B-type Natriuretic Peptide (BNP) levels (detect heart failure and elevate when CHF is worsening). Chest x-ray will detect any change in size of the heart. The EKG (Electrocardiogram) displays the electrical pulsed from the heart (see chapter 8 & 11). Ejection fraction (EF) will measure how well your heart is pumping blood. Additionally, cardiac stress test (see chapter 11) and cardiac catheterization (see chapter 11) will help to determine how well the heart is working.
TREATMENT	Your healthcare provider will monitory closely for changes in your condition, treat any hypertension (high blood pressure) and edema (water retention) with diuretics (water pills), angina (chest pain), or any underlying cause. Reduce your salt intake (which causes water retention or edema), check for rapid weight gain caused by water retention and take your medications as prescribed. Supplemental oxygen is almost always prescribed (at least for use at night). New treatment protocols are being developed every day. Adult stem cell therapy may be used to create new heart muscle and/or new blood vessels as an alternative to bypass surgery (see chapter 11 for bypass surgery).

Note: Heart failure causes your kidneys to function less efficiently due to decreased blood flow. This limits your kidneys ability to remove salt and excess fluid which causes edema (swelling).

Cor Pulmonale (Right Heart Failure)

DEFINITION	Right Ventricular enlargement secondary to (caused by) a malfunction of the lungs producing pulmonary artery hypertension (high arterial pressure in the lungs).

CAUSE | Sever diseases can lead to the respiratory failure (COPD, extensive loss of lung tissue, chronic pulmonary emboli, etc). Depending on the cause, it may be a reversible condition.

SYMPTOMS | Early in process little or no symptoms (asymptomatic). As it progresses substernal chest pain (angina), dyspnea (shortness of breath or difficulty breathing), venous distention (veins look swollen or puffy), fatigue with exertion, lightheadedness (dizziness) and/or fainting (with exercise or when standing up) may develop.

TESTS | Chest x-ray will reveal an enlarged right ventricle. An EKG (electrocardiogram) will show changes specific to this process will help in diagnoses and the heart's function can be evaluated with radionuclide studies (using an atom that disintegrates by emitting electromagnetic radiation), echocardiography (sonar image of the heart), and cardiac catheterization (see chapter 6, 8 & 11).

TREATMENT | To relieve right-sided heart failure, oxygen, diuretics (used to remove excess fluids) and anticoagulants (blood thinners) are prescribed to treat the underlying lung disease and resolve (eliminate or reduce) pulmonary hypertension (elevated blood pressure in lungs due to extra fluids).

Coronary Artery Disease (CAD)

DEFINITION | A disease caused by plaque build up inside arteries that supply oxygen and food to the heart muscle which cause heart attacks or congestive heart failure (CHF).

CAUSE | Arteries become narrowed or blocked by plaque reducing the blood flow to the heart muscles usually caused by cholesterol and high blood pressure. CAD, arrhythmias (abnormal heart beat) and nocturnal complex arrhythmias (abnormal heart beat at night) have been associated with Sleep Disordered Breathing (SDB) also known as Sleep Apnea.

SYMPTOMS | Can be asymptomatic (no symptoms), angina (chest pain) which is a sign to slow down, dyspnea (shortness of breath with exertion), rapid heart beat, nausea, fatigue (weakness), and dizziness. A very bad sign or symptom of CAD is a heart attack (see heart attack) and this may be the first indication you have a problem.

TESTS | Your healthcare provider will determine if your angina (chest pain) is stable (pain diminishes with medication or rest) or unstable angina (pain changes, lasts longer or does not diminish like normal with rest or medication). Laboratory blood tests will be ordered to determine if you have had a heart attack or to determine if other heart disease is present. EKG (electrocardiogram) will be administered to see if any damage to the heart has occurred and that the heart is working properly (see chapter 8 & 11). Additionally a MRI (Magnetic Resonance Imaging Scan), PET (Positron Emission Imaging Scan), or CT (Computed Tomography Scan) may help make a definitive diagnosis.

TREATMENT | For the most part treatment for angina (chest pain) is supportive to help you manage the pain. Your healthcare provider will treat hypertension (high blood pressure), edema (any retained fluids), cardiac catheterization (see chapter 11) to stent or remove the plaque or in some cased heart transplant may be needed (see chapter 11). New treatment protocols are being developed every day. Adult stem cell therapy may be used to create new heart muscle and/or new blood vessels as an alternative to bypass surgery (see chapter 11 for bypass surgery).

Coronary Heart Disease (see Coronary Artery Disease (CAD))

Coronaviruses (cause about 25% of colds), (see Rhinovirus)

Croup (Laryngotracheobronchitis or viral croup) PEDIATRIC

DEFINITION   A common viral inflammation of the upper and lower respiratory tracts resulting in airway obstruction.

CAUSE   Caused by Parainfluenza virus, RSV (Respiratory Syncytial Virus) or influenza A or B viruses.

SYMPTOMS   Barking cough, hoarseness, stridor (harsh high-pitched sound), fever (in 50% of the cases), intercostal retractions (use of chest muscles to assist in breathing), cyanosis (a blue tint to skin, lips or fingernails), shallow respirations (may develop), dyspnea (shortness of breath or difficulty breathing), and subglottic swelling (narrowing of the upper trachea). Breath sounds expiratory wheezes, rales/rhonchi (wet sounding lungs), and stridor (harsh high-pitched sound).

TESTS   X-ray to rule out Epiglottitis (see Epiglottitis) and confirm subglottic narrowing of trachea (Steep Sign). ABG (Arterial Blood Gas) reveals hypoxemia (low blood oxygen level) with or without hypercapnia (elevated carbon dioxide level).

TREATMENT   Humidification may help in mild cases however, depending on severity (i.e. breath sounds, cough, retractions (chest muscles being sucked in) and cyanosis (a blue tint to skin, lips or fingernails), hospitalization may be required. Care will include; cool mist, racemic epinephrine, dexamethazonem (steroid) and/or budesonide (steroid). Severe cases may require mechanical ventilation.

Cryptococcosis (Torulosis)

DEFINITION   Infectious disease due to the fungus Filobasidiella (fungus found in pigeon droppings). Its primary focus is in the lungs and can spread to the meninges (three layers of membranes that cover the spinal cord and brain), kidneys, bone and skin.

CAUSE   Infection acquired through inhalation of the fungus patients with immunosuppressive diseases (AIDS, Hodgkin's, etc.) are at highest risk.

SYMPTOMS   Primary focus is the lungs however; occasionally the kidneys, bone and skin are infected. Only recently has it been recognized that a benign, rarely progressive pulmonary form occurs and is often as a complication of other lung diseases. Cough or other cold like symptoms may be present however, the patient may be asymptomatic (no symptoms). Meningitis with headache is the most common symptom with blurred vision or mental disturbances (confusion, depression, agitation and/or impaired speech).

TESTS   Laboratory test for budding yeast surrounded by a clear capsular area in sputum, and cerebrospinal fluid (CSF). Additionally, CSF and blood are tested for antibodies. Chest x-ray may reveal infiltrates (junky looking lungs).

TREATMENT   In the non-progressive pulmonary disease treatment may not be needed. Other forms of the disease usually include antifungal medications.

Cryptococcosis Gattii (Cryptococcosis, Cryptococcus or Cryptococcal Disease)

DEFINITION   Potentially fatal airborne fungal lung disease that comes from soil and trees.

CAUSE   Inhalation of fungus. Patients with immune system compromise (reduce ability to fight off infections or weakened immune system), or taking high doses of corticosteroid medications, chemotherapy patients, transplant patients, Hodgkin's Disease and AIDS are most at risk however; due to recent mutations of the fungus even healthy people are at risk of infection (about 25% fatality rate).

SYMPTOMS   Chest pain, dry cough, fever, headache, fatigue, nausea, skin rash, swelling of the abdomen, swollen glands, weight loss, confusion and blurred vision.

TESTS   Laboratory testing of cultures for the antigen (is what your immune system recognizes) using blood, sputum, urine and cerebral spinal fluid are used for diagnosis. Additionally, your healthcare provider will order a chest x-ray and may order a bronchoscopy (see chapter 11) or a lung biopsy (remove a small piece of tissue from the lung for testing (see chapter 11).

TREATMENT   Treatment can last for months using IV medications combined with oral medications (i.e. Amphotericin B, flucytosine, fluconazole) may be prescribed.

Cystic Fibrosis (Mucoviscidosis, Pancreatic Enzyme Deficiency)

DEFINITION   An inherited disease of the exocrine glands which causes an excess amount of abnormal secretions, resulting in tissue and organ damage of the lungs and digestive tract.

CAUSE   Cystic fibrosis (CF) is the most common lethal genetic disease (inherited disease) caused by inheriting an autosomal recessive trait, which are mutations of the CF gene. Insufficiency in supply of trypsin and lipase and the control of the production of the protein that regulates transport of chloride and sodium across cell membranes is disrupted, resulting in multiple organs and glands being affected causing adverse effects in the lungs, pancreas, liver, and sweat glands.

SYMPTOMS   It is usually characterized by developing a combination of Chronic Obstructive Pulmonary Disease (COPD), exocrine pancreatic insufficiency and high sweat electrolytes (sodium chloride) concentrations. Developing a persistent cough with very large amounts of thick (viscous) mucus, dehydration (low fluid levels in the body) and, frequent bronchitis, pneumonia, bronchiectasis and/or atelectasis (airless portion of the lungs). Additionally, shortness of breath (dyspnea), chest congestion, cyanosis (a blue tint to skin, lips or fingernails), hypoxemia (low blood oxygen level), intestinal obstructions, digestive problems, nutritional deficiencies, and digital clubbing. Physical signs include swelling of the lymph nodes, blocked bile ducts in the liver, gallbladder obstruction, and pancreatic obstruction. Breath sounds reveal rhonchi and wheezes (see chapter 8).

TESTS   Pilocarpine iontophoresis sweat test (collect sweat to identify electrolyte concentrations). Pulmonary Function Test (PFT) will reveal an obstructive disease pattern.

TREATMENT   Routine Mucolytics (mucus thinning), bronchial dilator, corticosteroids and antibiotics, all given via nebulizer treatments combined with an aggressive bronchial hygiene (moving secretions), secretion control and clearance regiment using a Therapy Vibratory Vest (vibrating vest, looks like a life preserver), Chest Physical Therapy (CPT), and/or vibratory PEP will ease the work of breathing.

Deep Vein Thrombosis or DVT

DEFINITION   A form of thrombophlebitis (inflammation of a vein with the formation of a clot) normally in the leg or pelvic veins although it can form in other areas.

CAUSE   Decreased blood flow rate, damage to the blood vessel wall and hypercoagulability (rapid or increased clotting). Medical conditions such as vein compression, trauma (damage), cancer, some inflammatory diseases (stroke, heart failure or nephritic syndrome) and infections can lead to DVT formation. Your risk of DVT formation is increased with surgery, casts, immobilization, smoking, age, obesity, and some medications.

SYMPTOMS   Asymptomatic (showing no symptoms) to extremity (arm/leg) pain, swelling, redness, warm and the superficial veins (ones you can see) are swollen (engorged). If a clot dislodges (breaks free into the blood) you have a medical emergency. If it were to lodge in the lungs it will cause a Pulmonary Embolism (PE) resulting in dyspnea (difficult breathing), Tachypnea (rapid breathing), and tachycardia (increased heart rate) and in severe cases death. A late complication is post phlebitic syndrome which causes edema (water retention and swelling), pain, muscle/skin discomfort and skin problems.

TESTS   The gold standard for testing is the intravenous venography which is an x-ray using a contrast agent (like a dye to see the obstruction better). The laboratory test called D-dimers and a Doppler Ultrasound are most often ordered to diagnose DVT.

TREATMENT   Prevention is the best bet, exercise your extremities often when traveling and wear those ugly compression stockings after surgery. Your healthcare provider may prescribe anticoagulation medications (blood thinners), thrombolytic (clot busters) or surgery in severe cases.

Note: Hormone therapy may increase your risk of DVT formation.

Diabetes Mellitus (DM)

DEFINITION   A syndrome resulting from hereditary and environmental factors and is characterized by abnormal insulin secretion, elevated blood glucose (sugar) levels and organ complications.

127

CAUSE	The inability of the body to control its sugar level. If insulin is not available to convert the sugar to food within the blood the production of keto acids (basically carbon) and acetone are increased and this decreases the blood's pH (low pH is acidosis). Type 2 diabetes has been associated with Sleep Disordered Breathing (SDB) also know as Sleep Apnea
SYMPTOMS	Most patients experience dizziness, confusion or coma. Additionally, if metabolic acidosis (increased acid levels not due to respiratory) is present the patient will experience headache, air hunger (severe hyperventilation (deep and rapid) or Kussmaul breathing), fruity acetone odor on the breath (almost a sickening sweet smell), nausea/vomiting, abdominal tenderness, extreme thrush (small white bumps on the tongue), and dry mucous membranes. Diabetes doubles your chances of heart attack and stroke.
TESTS	Blood draws for the laboratory to check glucose level, ketones and other markers. Arterial Blood Gas (ABG) is ordered to determine the acid/base balance. Personal glucose (sugar) monitors are of tremendous benefit in managing your sugar level.
TREATMENT	Depends on the type and severity, diet and medications can help control symptoms.

Note: Children who are obese are more likely to develop diabetes.

Diastolic Heart Failure (DHF)

DEFINITION	The abnormality in the heart's left ventricle filling during diastole (relaxed not beating).
CAUSE	The deterioration of preload compliance and E:A Ratio causing blood to regurgitate back into the left atrium and toward the lungs. This is normally due to the wall stiffness or thickness of the ventricle.
SYMPTOMS	Symptoms include hypertension (high blood pressure), aortic stenosis (stiffening or constriction) diabetes, heart failure, dyspnea (difficulty breathing, fatigue and/or chest pain.
TESTS	Diagnosis is difficult and imprecise. Echocardiography (sonogram, CT, PET, or MRI of the heart), Ejection Fraction, the decline of injection fraction compared with the E:A ratio can be used for diagnosis in some cases.
TREATMENT	Since it is normally a chronic process and can be well tolerated no specific treatment may be required. A surgical procedure exists however, it is not approved by the FDA yet. Medications such as calcium channel blockers may be beneficial. Pulmonary edema (extra fluids in the lung tissue) due to back flow from the heart to the lungs using diuretics may be prescribed.

Ductus Arteriosus (Fetus/Neonate)

DEFINITION	A channel for blood flow from the main pulmonary artery and the aorta (on the aortic arch) in the fetus.
CAUSE	While the lungs are full of fluid it protects the lungs from being overworked and allows the right ventricle to strengthen. This pathway

should close after birth.

Dust Pneumonia (Brown Plague) (see Silicosis, Black Lung Disease and Pneumonitis)

DEFINITION   Excessive exposure to dust (such as in dust storms) which overwhelms the lungs with dust (stopping the cilia or little hairs from removing the dust from the lungs) and inflaming the alveoli (the tiny air sacs where oxygen is exchanged in the blood) resulting in pneumonia or silicosis.

CAUSE   During the mid-1930's and around the world today (i.e. China, Africa etc.) the extreme lack of rain causes top soil and fine sand to be picked up by the wind and can be carried for hundreds of miles in dust storms.

SYMPTOMS   Common symptoms are coughing, high fever, difficulty breathing (dyspnea), nausea, chest pain, eye infections and laryngitis. People with asthma, children and the elderly are at the greatest risk.

TREATMENT   Prevention is the best bet by avoiding these dust storms or using surgical masks however they will become plugged with dirt is as little as one hour. Your home should be dust tight as possible.

DVT (see Deep Vein Thrombosis)

Embolic Stroke (see Stroke)

Emphysema

DEFINITION   An enlargement of the air spaces distal (past) to the terminal non-respiratory bronchioles (deep in the lungs) with destructive changes of the alveolar walls (see chapter 7). Also known as a "pink puffer" and is classified as COPD.

Note: Subcutaneous emphysema is easy to spot, by looking for the presence of large air-filled spaces (bubbles) in the soft tissues. You can actually feel and see it under the skin (feels like puffy or crispy rice).

CAUSE   The distention (pushing outward) of interstitial (internal) tissues of the lungs usually caused by smoking, air pollutants, occupational exposure to inhaled chemicals and aging. Additionally, emphysema can be caused by an alpha antitrypsin (a protein) deficiency or genetic emphysema.

SYMPTOMS   Since the alveoli (air sack) walls are damaged or destroyed, old air can not get out (air trapping) and new air can not get in. Increased airway resistance, with air trapping that results in chronic bronchitis, and frequent infections. Persistent productive cough (mucus production), dyspnea (shortness of breath or difficulty breathing), cyanosis (a blue tint to skin, lips or fingernails) and orthopnea (difficult breathing if not setting upright or standing) are present. Additionally, increased use of accessory muscles (chest muscles), barrel chest (chest is actually shaped like a barrel), labored breathing, pursed lip breathing (looks like they are trying to whistle on exhalation), digital clubbing (side to side widening of finger/toes and curvature of the nails) late forming symptoms. Many patients experience weight loss, and engorged neck veins (neck veins look swollen).

TESTS	Breath sounds reveal diminished flow, rhonchi (wet sounding) and wheezes (see chapter 8). Arterial Blood Gas (ABG) will reveal low blood oxygen (hypoxemia), hypercapnia (elevated carbon dioxide level), fully or partially compensated respiratory acidosis (where the body used bicarbonate and breathing rate to help control acid base balance). Pulmonary Function Test (PFT) reveals an obstructive disease (see PFT in chapter 11), and a decreased DLCO (diffusing capacity). The chest x-ray will reveal hyperinflation (over inflation) of the lungs, increased chest diameter (barrel chest), and a depressed diaphragm.
TREATMENT	Patient education is essential to learn to avoid pollutants and chemicals. Pulmonary rehabilitation and nutrition training are very valuable in maintaining health and quality of life. Smoking cessation education will make quitting smoking much easier (see Smoking Cessation at the end of this chapter). Most patients will use low concentrations of oxygen <4 lpm (protecting the hypoxic drive see note below). Medications can include bronchodilators, steroids, and expectorants (used to hydrate sputum and help you cough up the sputum). Surgical procedures considered will be lung resection to limit lung volume and lung transplant by your healthcare provider (see chapter 11).

NOTE: The hypoxic drive (backup system to trigger a breath) is the body sensing low oxygen levels to stimulating breathing. Normally the body will sense rising levels of carbon dioxide to stimulate breathing. However, people with chronic lung problems may loose the ability to sense the carbon dioxide. If too much supplemental oxygen is given to the patient, he/she may stop breathing due to high levels of oxygen in the blood.

Emphysema Neonate (see Pulmonary Air-Block Syndrome)

Empyema (also see Pleural Effusion)

DEFINITION	The Presence of purulent fluid (pus) in the pleural space (between lungs and chest wall).
CAUSE	Pneumococcal or Tuberculosis infections.
SYMPTOMS	Depending on severity, symptoms include; difficulty in breathing (dyspnea), tachypnea (rapid breathing >20), chest pain, cough, hemoptysis (bleeding in lungs), and hypoxemia (low blood oxygen level).
TESTS	The chest x-ray reveals a radiopacity (cloudy appearance) over the area where the fluid has collected. Breath sounds will reveal diminished sounds (over the affected area), and/or crackles (see chapter 8).
TREATMENT	Some cases may resolve with just antibiotics to treat the infection. More severe cases may require the placement of a chest tube into the chest wall (see chapter 6 & 11) for drainage (for months in some cases). The administration of supplemental oxygen therapy as required.

Endocarditis

DEFINITION	It is classified in multiple ways but is basically the inflammation of any internal structure or lining of the heart (i.e. endocardium, heart valves, interventricular septum, etc.).

CAUSE	There are many pathogens that can cause Endocarditis. Patients, who have a damaged heart valve, have an artificial heart valve or certain types of heart murmurs have an increased risk of developing this infection. It can be caused by a microorganism or bacteria (infective) and fibrin or platelets (non-infective) attaching themselves to a damaged area of the heart.
SYMPTOMS	Symptoms can be rapid onset (acute) or develop slowly over time and will be linked to the severity of the infection. Symptoms may include; fever, chills, fatigue, joint or muscle pain, shortness of breath (dyspnea), night sweats, cough, paleness, weight loss, and/or swelling of the legs or abdomen (edema). In addition tenderness on the left side under the rib cage, heart failure, kidney failure, stroke or septicemia (blood poisoning), small red spots in the whites of your eyes or skin (called petechiae).
TESTS	Since the condition is rare in people with healthy hearts, a good medical history is important for determining what may have damaged your heart. Blood tests will include blood cultures (to identify the bacteria present). An EKG will be ordered in order to detect an irregular heartbeat or electrical activity. A chest x-ray may reveal any damage or spread of the infection and may be followed by a CT scan to detect any spread of the infection into other organs. A urine sample may reveal blood in your urine. A transesophageal echocardiogram (see chapter 11) or ultrasound may be ordered to get a close up view of the heart valves.
TREATMENT	If the cause is bacterial your healthcare provider will order intravenous (IV) antibiotics followed by oral antibiotics which can last six weeks or longer (maybe over your lifetime). In cases where damage has occurred within the heart you may require surgery to repair the damage (see chapter 11 for the procedure). Prevention is important; dental hygiene (brush and floss), avoid skin infections, and discuss susceptibility to Endocarditis with all of your caregivers.

Epiglottitis (Supraglottitis) PEDIATRIC

DEFINITION	Acute severe, rapidly progressive infection of the epiglottis (see chapter 7) and surrounding tissue, often becoming life threatening due to sudden airway obstruction.
CAUSE	The Hemophilus influenza Type B infection is usually the pathogen in children 2-5 years old and is very uncommon in children under 2 years.
SYMPTOMS	High fever, stridor (harsh high-pitched sound), hoarseness, sore throat, labored breathing with a muffled voice and difficulty swallowing or drooling are common.
TEST	The lack of a barking cough helps differentiate from croup. X-ray shows markedly enlarged epiglottis (beyond the tongue and above the larynx (voice box)) and distention of the hypopharynx (just above the epiglottis).
TREATMENT	Rapid treatment is essential. Some severe cases require sedation and nasotracheal intubation (ET tube inserted through the nose into the trachea (see chapter 6 & 11)) or tracheostomy (hole with breathing tube placed through the skin on the neck (see chapter 6 & 11)). Nasotracheal intubation should only be for 24 to 48 hours.

Fibrogenic Dust Disease (see Silicosis, Asbestosis or Coal Workers Pneumoconiosis)

Fibrosis, Pulmonary

DEFINITION
Fiber replacement of the normal tissue or the lungs, organs and/or other structures the body.

CAUSE
The cause is not defined in 50% of the cases (Idiopathic Pulmonary Fibrosis) and some cases is believed to be due to inhalation of mineral dust like silica (Silicosis) or sand. Additionally, it can be caused by oxygen toxicity (high levels of supplemental oxygen over a long period of time), necrosis (dead tissue), infection, breathing noxious gases, pneumonitis, sarcoidosis, ARDS, aspiration (sucking foreign matter into lungs), and/or bronchiectasis.

SYMPTOMS
Pulmonary fibrosis replaces or obstructs the tissue and reduces or stops the normal function of the lungs. Symptoms may include shortness of breath (dyspnea), increased work of breathing, decreased lung compliance (the lung looses elasticity) and hypoxemia (low blood oxygen level). Breathe sounds will be decreased (diminished) with rales (wet sounding) over affected area.

TESTS
X-ray shows fibrous areas (foggy white), atelectasis (airless portions of the lungs), and a possible mediastinal shift (movement of the lungs and trachea) toward fibrous area.

TREATMENT
Treat infections and complications. Supportive care with supplemental oxygen. Patient education is avoiding exposure to pulmonary irritants is critical.

Fixed Obstructive Lung Disease (see Popcorn Workers Lung)

Flail Chest

DEFINITION
The fracture of two or more ribs, in more than one location, resulting in unstable area of the chest wall caused by trauma. The unstable portion of chest wall moves in opposite direction (paradoxical) of the normal movement of the chest wall.

CAUSE
Extremely painful making it difficult to breathe. A collection of air (pneumothorax) or fluids (hemothorax) between the lungs and the chest wall at the injury site often results.

SYMPTOMS
The paradoxical respirations causes intense pain in affected area, dyspnea (shortness of breath or difficulty breathing) and increased work of breathing,

TESTS
Chest x-ray reveals fractured ribs and possibly atelectasis (white plate like or ground glass) at the site of the injury. ABG (Arterial Blood Gas) will reveal hypoxemia (low blood oxygen level).

TREATMENT
Stabilize chest wall, chest tube may be needed for hemothorax or pneumothorax (fluid or air between the chest wall and lungs) to relieve the pressure, supplemental oxygen as required. In extreme cases, mechanical ventilation may be required due to respiratory failure.

Foramen Ovale (PFO), (Fetus/Neonate)

DEFINITION   An opening between the right and left atrium (upper changes of the heart). This opening usually closes within the first year of birth.

Gastroesophageal Reflux Disease (GERD)

DEFINITION   A chronic digestive disease occurring when stomach acid or bile (backup) forces stomach content into the esophagus (Food Pipe)and in some case the trachea (wend Pipe).

CAUSE   Stomach contents being forced back into the esophagus. GERD has been associated with Sleep Disordered Breathing or Sleep Apnea.

SYMPTOMS   Acid reflux and heartburn more than 2 time per week causing a burning sensation in your throat, chest (heartburn), bad taste, chest pain, dysphagia (difficulty swallowing, lump in your throat, dry cough, sore throat and/or hoarseness. Pregnancy, obesity, smoking, hiatal hernia, connective tissue disorders, diabetes and asthma increase you risk of developing GERD. Over time narrowing of the esophagus (esophageal stricture), precancerous changes in the esophagus (Barrett's Esophagus) and open sores in the esophagus (esophageal Ulcer) will develop. If the stomach contents is aspirated (inhaled into the lungs) the lungs will develop aspiration associated diseases (such as pneumonia). Associated with Sleep Apnea or Sleep Disordered Breathing (SDB) due to the negative intrathoracic pressure (inside the chest area) caused by closed airways.

TESTS   X-ray of the upper digestive system (barium swallow or upper GI series) in order to detect hollow areas of the digestive tract, esophagus and upper intestine (duodenum). Your health care provider may order Endoscopy (small tube passed through your stomach) in order to visualize or take tissue samples from the esophagus and stomach. A pH and impedance probe is passed into your esophagus to test acid levels and movement.

TREATMENT   Over the counter medications such as antacids (neutralizes acid) and H2 receptor blockers (blocks acid production are often beneficial. Prescription medication called proton pump inhibitors may be required. In sever cases surgery may be required to tighten the lower esophageal sphincter preventing reflux, surgery to create a barrier preventing the stomach acid backup or surgery to form scar tissue in the esophagus.

Note: Seek immediate medical care (call 911) if you develop chest pain, shortness of breath (dyspnea), jaw pain, arm pain and/or any other symptoms of a heart attack.

Gastroesophageal Reflux Disease (GERD) PEDIATRIC - CHILD - ADULT

DEFINITION   The regurgitation (vomiting) of stomach contents into the esophagus, which could be aspirated (sucked into the lungs).

CAUSE   Caused by physiological defect (defective or weak) or damage to the esophageal sphincter (valve at the top of the stomach).

SYMPTOMS   The development of Respiratory Reactive Airway Disease, aspiration pneumonia (caused by sucking stomach contents into the lungs), laryngospasm (muscle spasm of the voice box), stridor (harsh high pitched sound), chronic cough, choking spells and/or apnea (cessation of breathing). This could be an acute (sudden onset) life-threatening event for an infant on an indication when an older child has unexplained chronic head and neck problems.

TEST   Laboratory esophageal pH testing (tests for acid presence), Upper GI (gastrointestinal) contrast study, and gastric scintiscan (special scan used with radioactive substance for scintiphotography).

TREATMENT   Management is normally through H2 blockers or proton pump inhibiters (medication that reduces the secretion of stomach acid) or surgical intervention if other management is not effective.

Goodpasture's Syndrome

DEFINITION   An uncommon autoimmune disorder causing bleeding in the lungs (hemoptysis, coughing up blood) and progressive kidney failure.

CAUSE   Autoimmune system disorder.

DEFINITION   A group of several disorders that affect the small spaces of the lungs.

Guillain-Barre' Syndrome

DEFINITION   An idiopathic (no known cause), peripheral polyneuritis (multiple muscle paralysis) characterized by lower extremity weakness that progresses to the upper extremities and face. May lead to flaccid paraplegia (loss of control or muscles) as well as marked respiratory muscle weakness.

CAUSE   Thought to be caused by an autoimmune reaction (where the body attacks itself) in which the body's immune system attacks the myelin sheath (protective layer around nerves) it usually occurs five days to three weeks after an infection (flu-like), surgery, or immunization.

SYMPTOMS   Ascending (starting with the legs and moving upward) muscle weakness and in acute (rapid onset) cases can become a medical emergency, although most are mild cases and can resolve on it's own over several months. Symptoms can include pin like sensations, numbness, and burning pain, to a total loss of feeling and/or control. Progression of the paralysis can often lead to muscle paralysis (inability to move muscles), including the diaphragm and muscles of ventilation resulting in respiratory failure.

TESTS   An electromyography (study using electrical stimulation of muscles), and nerve conduction studies are uses to test movement and sensation. Laboratory analysis of cerebrospinal fluid for high protein levels and blood/urine testing are used to detect or rule out other problems (i.e. thyroid disorders, diabetes, kidney failure, etc.). Vital Capacity (VC, which is the total amount of air that can be exhaled after a maximum inspiration) is checked frequently (see PFT in chapter 11) at the patent's bedside.

TREATMENT   The patient's airway must be kept unobstructed (clear) and their ability to breathe must be measured frequently in order to determine the progression or regression of the paralysis (loss of muscle control). Care is mostly supportive; treat symptoms or complications, ensure nutritional support, physical therapy is ordered as required. Plasmapheresis is a laboratory procedure, which removes autoanibodies and proteins then replaces the blood. This can be a short-term illness or a permanent condition, which may require ventilator support.

H1N1 Or H1N3 (see north-American Influenza, Swine Flu)

H5N1 (see Avian Flue, Bird Flue)

Heart Attack (see Myocardial Infarction)

Heart Disease (Cardiopathy) (see specific disease)

DEFINITION   It is many different diseases affecting the heart which are the leading cause for death in America.

Heart Failure (see Congestive Heart Failure or CHF, Cor Pulmonale)

Heart Palpitations (arrhythmias)

DEFINITION   When you feel your heart start beating too fast (tachycardia), too slow (bradycardia), too hard (bounding feeling) or just beats irregularly.
CAUSE   There are many causes other than heart disease (i.e. fear, caffeine, stress etc.) however, it could be a wakeup call to discuss it with your healthcare provider.

Heart Tumors

DEFINITION   These tumors are abnormal growth of heart tissue.
CAUSE   There are primary heart tumors (from heart tissue) and secondary heart tumors (come from another part of the body).
SYMPTOMS   Symptoms can be asymptomatic (without symptoms), arrhythmias (abnormal heart beat), decreased blood pressure (hypotension), heart murmurs (sounds caused by abnormal blood flow within the heart) and/or blood clots (caused by pieces breaking off and becoming stuck anywhere in the body).
TESTS   Primary heart tumors may be very difficult to diagnose and secondary heart tumors may be present in cancer patients. An EKG may detect changes (from previous EKG traces), heart murmurs can be heard while listening to the heart (auscultation), echocardiography and ultrasound (to detect blood flow obstructions or limitations), CT, MRI, chest x-ray or Transesophageal echocardiography or Coronary Angiography (to detect structural changes of the heart and blood vessels). If a tumor is found a cardiac catheterization is performed to biopsy (obtain a piece for testing) is performed (see chapter 11).
TREATMENT   Many tumors do not require treatment. Primary heart tumors which are cancerous can not be removed (because they will spread throughout the body) and are in most cases fatal. Radiation or chemotherapy may be prescribed. Depending on the location and their affect on ht heart,

primary heart tumors which are noncancerous will be removed surgically and have a positive prognosis (outcome).

Note: Most heart tumors are benign (not malignant) but can be just as deadly if they restrict the heart from beating or pumping properly. Additionally, if a piece should break off the tumor it can be carried around the body causing a stroke or clot depending on its final resting location.

Heart Valve Defects (see Congenital Heart Disease NEONATAL)

Heart Valve (General View of its Dysfunction)

DEFINITION	They do not open enough or cause a reduced flow (stenosis) or do not close properly (leak, insufficiency or regurgitation).
CAUSE	Can be caused by Rheumatic Fever, calcification (calcium buildup on valve), medications, or an unrepaired small birth defect. Over time the birth defect will worsen causing scarring or even destroy a valve.
SYMPTOMS	The heart must work harder which can lead to heart failure or Congestive Heart Failure (CHF). A mitral Valve Prolapse is diagnosed when the valve does not close properly (more common in women). This causes an increased pressure in the left ventricle and pushes the Mitral Valve back inside the left atrium causing a leak. Patents with Mitral Valve Prolapse are at high risk for endocarditis which is an infection within the heart (see Endocarditis). The patient may be asymptomatic (no physical signs of problems) or the reduced pumping ability causes dizziness, dyspnea (shortness of breath) and poor peripheral perfusion (supply of blood to the body).
TESTS	A Doppler of the heart will show the leak and heart sounds a changed by ether leaks or stenosis (narrowing). The Ejection Fraction (EF) or the blood the heart is able to pump may be greatly reduced.
TREATMENT	Mitral Valve Prolapse may require life-long use of antibiotics and your healthcare provider will prescribe antibiotics anytime you are having any invasive procedures (anything that makes a hole in your body, like dental procedures or minor surgery). Some patients with Mitral Valve Prolapse will never require repair or replacement however surgical repair or replacement may be necessary (see chapter 11).

Hemorrhagic Stroke (see Stroke)

Hemothorax

DEFINITION	Presence of blood or fluid in the plural space between the chest wall and lung.
CAUSE	Most follow chest trauma or injury, however they can occur spontaneously (just happens).
SYMPTOMS	Depending on the size of the fluid area symptoms will vary from asymptomatic (no symptoms) to severely compromised ventilation (difficulty breathing or inability to breathe) due to the fluid taking up space placing pressure on the lungs and not allowing the lungs to inflate or expand correctly. Symptoms include; dyspnea (shortness of breath or

difficulty breathing), tachypnea (rapid breathing >20) and tachycardia (rapid heart rate >100) and increased work of breathing (tires the patient out).

TESTS	Chest x-ray will indicate the affected area. Percussion (tapping on the chest while listening to the thumping noise (see chapter 8)) is dull over affected area and breath sounds will be decreased or absent. If the area is large enough the trachea will pushed away from affected side, (the fluid pushes the lungs over).
TREATMENT	This is often a medical emergency requiring trauma room care. Chest tube drainage (see chapter 6 & 11) is inserted between the chest wall and lung to remove the excess fluid however; if the area is small enough, a needle may be inserted to remove (drain) any fluids (called thoracentesis, see chapter 11). Supplemental oxygen is given as required and antibiotics are used to treat any infections.

Note: Pleural blood often does not clot well making bleeding hard to stop.

Histoplasmosis

DEFINITION	A rare disease caused by inhalation of fungal spores, which initially causes an infection in the respiratory, and GI tract (digestive tract).
CAUSE	After inhalation of dust containing the spores, an infection develops causing calcifications within the lungs and other organs.
SYMPTOMS	Vary from asymptomatic (no symptoms) to becoming fatal. The spores cause ulcerations of the oropharynx (back of the throat), GI tracts (digestive tract) and skin. Often fever, cough, malaise (fatigue), flu-like symptoms, dyspnea (shortness of breath or difficulty breathing) and/or chest pain.
TESTS	Chest x-ray may reveal calcified pulmonary lesions (white spots). Laboratory testing of tissue samples are microscopically analyzed after staining (a procedure to identify the disorder). Blood tests are not conclusive until late in the disease process.
TREATMENT	Amphotericin B is administered and chemotherapy is rarely required. It is fatal only in those rare untreated cases when a major infection is present at the same time.

HIV (see Acquired Immune System Disorder)

Note: There are now four strains of HIV virus. Three related to the viruses in simian or chimpanzee and the fourth strain HIV-1 related to viruses in gorillas.

Hyper-reactive Airways (see Asthma)

Hypertension (High Blood Pressure)

DEFINITION	An abnormally high blood pressure (see chart below) which is normally chronic (ongoing). Determined by the quantity of blood pumped by the heart and the resistance in the arterial walls (see table below).
CAUSE	Classified by essential (primary) which has no cause or secondary (about 5% to 10% of hypertension cases) which is caused by another condition

are normally a complex chronic medical condition (sleep disturbances, kidney disease, adrenal abnormalities, pregnancy, medications and others). It can also be caused by diet, stress, obesity, alcohol, tobacco and others. Additionally, it has been associated with Sleep Disordered Breathing (SDB) also known as Sleep Apnea.

SYMPTOMS   Mild to moderate essential hypertension is normally asymptomatic (no symptoms). Symptoms can include nose bleeds, nausea, vomiting, headache, seizure, fatigue, dyspnea (difficulty breathing), somnolence, confusion, and/or visual disturbances. Risk factors include stroke, heart attack, heart failure, aneurysm (swollen blood vessels), renal failure (Kidneys) and decreased life expectancy.

TESTS   Your healthcare provider will complete a detailed medical history and physical exam making his/her diagnosis on the basis of a persistently high blood pressure (over a 1 week period). Laboratory test will be used to identify reversible (secondary) causes (such as renal disease) and for organs that may have suffered damage (i.e. heart, eyes, kidneys, etc.). Additionally your healthcare provider will monitor you for diabetes, elevated cholesterol, cardiovascular disease and others due to your risk factors.

TREATMENT   Prevention is your best bet through eating correctly, weight reduction reducing stress and exercise. Treatment begins with what you should have done to help prevent hypertension and will add dietary changes beneficial to reducing blood pressure, reducing salt/sugar intake, discontinuing tobacco/alcohol use, biofeedback and significant lifestyle changes. Medications include antihypertensive drugs (to lower blood pressure), Diuretics (to help remove excess fluids), ACE inhibitors, Calcium Channel Blockers, Alpha Blockers and Beta Blockers most of which to manage the secondary and risk factor problems.

Classification of Blood Pressure		
Category	Systolic	Diastolic
	mmHg	mmHg
	Top #	Bottom #
Hypotension	<90	<60
Normal	90-119	60-79
Pre-Hypertension	120-139	80-89
Stage 1 Hypertension	140-159	90-99
Stage 2 Hypertension	≥160	≥100

Note: Children who are obese are more likely to develop hypertension.

Hypertensive Heart Disease

DEFINITION   Heart disease caused by high blood pressure.

CAUSE	Localized high blood pressure caused by congestive heart failure, enlarging of the left ventricle (Left Ventricular Hypertrophy), coronary heart disease and abnormal heart beats (cardiac arrhythmias).
SYMPTOMS	Irregular pulse (arrhythmias), chest pain, fatigue, swelling of the feet or ankles (edema), weight gain, shortness of breath (dyspnea), nausea and difficulty lying down to sleep. Potentially develops into Hypertensive Cardiomyopathy, Coronary Heart Disease, Left Ventricular Hypertrophy).
TESTS	Monitor blood pressure and EKG to check for damage.
TREATMENT	Prevention is your best bet. Treat the high blood pressure (i.e. diet, exercise, medications etc.), water pills (diuretics) to reduce swelling (edema), control diabetes, reduce cholesterol (hyperlipidemia), serous lifestyle changes, reduce alcohol use, loose weight and stop smoking.

Hypersensitivity Lung Disease (Allergic Pulmonary Diseases, such as Hypersensitivity Pneumonitis, Eosinophilic Pneumonias, Allergic Bronchopulmonary Aspergillosis)

Note: This is a very complicated disease. As you read the "CAUSE" below, we used the work "like" to give you an example of the types of reactions that cause the disease.

DEFINITION	One of or, a mixture of the four types of over sensitivities that cause several allergic lung diseases.
CAUSE	Allergic reaction to antigens (like the reaction in asthma), Cytotoxic reactions (like Goodpasture's Syndrome), Immune complex (like Systemic Lupus), and/or Cell mediated (like the positive reaction from a TB test).
SYMPTOMS	The symptoms are cause dependant and can range from shortness or breath (dyspnea) to coughing up blood (hemoptysis), fever, chills, wheezing and/or rales (breath sounds over the affected area, see chapter 8) , fatigue, weight loss, and/or nausea.
TESTS	Chest x-ray or CT scan to determine the affected area and damage. Blood and sputum testing to help determine the cause and required treatments. Pulmonary Function Tests (PFT) to determine lung function.
TREATMENT	Treatment will be dependant on cause and any other conditions associated with the disease.

Hypotension (Low Blood Pressure)

DEFINITION	Abnormally low blood pressure which can be life-threatening (see table below).
CAUSE	Hypovolemia (low blood volume) due to blood loss, insufficient fluid intake (dehydration), diarrhea, vomiting and excessive use of diuretics (water pills). Decreased cardiac output (not pumping enough) due to congestive heart failure, myocardial infarction (heart attack), bradycardia (slow heart rate <60) can progress into cardiogenic shock and arrhythmias (unusual heart beats) often result in hypotension (usually temporary). Some syndromes such as orthostatic hypotension (blood pressure changes based on the position of your body) and Neurocardiogenic Syncope (reduced blood pressure while standing which causes you to pass out) can also cause hypotension.

SYMPTOMS   Symptoms are widely varied and usually associated with other
                conditions, but may include syncope (passing out), seizure headache,
                fatigue (tiredness), blurred vision, nausea, chest pain, dyspnea (shortness
                of breath), foul smelling urine, and others.

TESTS   Laboratory blood tests for glucose (sugar), anemia (reduced red blood
                cells), and blood testing electrolytes (sodium, potassium, etc.). An EKG
                to detect electrical pathway problems with the heart and Holter or event
                monitors to observe heart activity over a longer period of time.
                Echocardiography (ultrasound of the heart) tests how well the heart is
                working and used to visualize (with moving pictures) its anatomy. A
                cardiac stress test will enable your healthcare provider to see how well
                your heart works under a higher workload. A Valsalva maneuver test the
                nervous system function through taking several deep breaths and
                exhaling through pursed lips and a Tilt Table Test where you lie on a
                table and the head of the table is tilted down then up. If you faint during
                the test, your healthcare provider may be able to diagnose orthostatic
                hypotension, neurally mediated hypotension or other brain or nerve
                conditions (such as the vagus nerve, see Valsalva Maneuver).

TREATMENT   Treatment will be dependant on the diagnosis, cause, signs and
                symptoms.

Classification of Blood Pressure		
Category	**Systolic**	**Diastolic**
	mmHg	mmHg
	Top #	Bottom #
Hypotension	<90	<60
Normal	90-119	60-79
Pre-Hypertension	120-139	80-89
Stage 1 Hypertension	140-159	90-99
Stage 2 Hypertension	$\geq$160	$\geq$100

Note: Children who are obese are more likely to develop hypertension.

Hypoxemia

DEFINITION   An abnormally low arterial blood oxygen level.

Note: Clinically an arterial oxygen level of $\geq$80 mmHg at sea level is normal.

CAUSE   Your body likes everything just right, in medical terms we call it
                homeostasis. There are to many causes to mention, but it's safe to say if
                your oxygen level drops below what you body needs bad things are
                about to happen.

SYMPTOMS   Shortness of breath or difficulty breathing (dyspnea), increased heart rate
                (tachycardia >100), the heart beats harder (bounding pulse), confusion,
                seizures and/or death.

TESTS	Pulse oximetry (the finger probe that glows red) is a routine rapid test to ensure adequate oxygenation. Arterial Blood Gas (ABG) is the most accurate and requires arterial blood for testing. Laboratory testing of blood is done to rule out other processes and to help identify the cause of the hypoxemia (low blood oxygen level).
TREATMENT	Supplemental oxygen will increase your blood oxygen level. Your healthcare provider may order oxygen while sleeping, during exercise or continuously. If possible, your healthcare provider will identify and treat any underlying conditions causing the deficiency.

NOTE: The degeneration of you lungs is a natural part of aging. It is not necessary to complete an ABG on room air however; many physicians believe it is necessary for a baseline.

Idiopathic Infiltrative Lung Disease (see Interstitial Lung Disease)

Inflammatory Heart Disease (Cardiomegaly)

DEFINITION	Swelling or inflammation of heart muscle and/or the tissue around the heart.
CAUSE	Inflammation of the inner layer of the heart (endocarditis), enlarged heart (cardiomegaly), inflammation or the heart muscle (myocarditis) and the most common is the heart valves.
SYMPTOMS	Shortness of breath (dyspnea), dizziness (light headed), fatigue, chest pain, and heart palpitations (arrhythmias).
TESTS	Chest x-ray (for sizing the heart), EKG to check for damage and laboratory blood work (testing for enzymes, electrolytes (i.e. sodium, potassium etc.).
TREATMENT	Prevention is your best bet. Treat the high blood pressure (i.e. diet, exercise, medications etc.), water pills (diuretics) to reduce swelling (edema), control diabetes, reduce cholesterol (hyperlipidemia), serous lifestyle changes, reduce alcohol use, loose weight and stop smoking.

Influenza

DEFINITION	An acute (rapid onset), usually self-limiting infectious viral disorder.
CAUSE	The flu is caused by the influenza virus which is divided into type "A", "B" or "C", each with several subtypes or strains. The viruses can be air-borne, can live for prolonged periods on anything you or your family can touch, and transmitted through direct physical contact. If you touch a contaminated surface, then touch your mouth, your nose, your food, or anything you put in your mouth or nose and you got it…
SYMPTOMS	Symptoms develop within the first 24 hours or so, beginning with chills and fever (102-103 F). These symptoms will be followed by sore throat, myalgia (tenderness/pain in muscles or aches/pains), headache, photophobia (over sensitivity to light), substernal burning (due to the coughing), nonproductive cough (no sputum), and malaise (discomfort or uneasiness). The cough will become productive (produce mucus), skin will be warm and flushed, reddened throat, watery eyes then subside very quickly.

TESTS	Due to the influenza type "A", "B" and "C" all having several subtypes or strains and antigenic drifts (mutant changes in the virus about every 10 years or less) laboratory testing will always lag behind. However, laboratory testing is done to rule out and treat any bacterial infections present.
TREATMENT	Prevention is the best bet. Flu vaccines are offered to cover expected outbreaks of specific viruses, however immunity would imply the virus will never change and this is not be the case. <u>Antibiotics are prescribed for bacterial infections only; they do not cure viral infections.</u> This is why your healthcare provider may send you home with only cough syrup and Nonsteroidal Anti-inflammatory Drugs (NSAID's). .

Insomnia

DEFINITION	A serious disease which causes the inability to get the required nights restful sleep (at least six hours).
CAUSE	It is difficult to determine a cause.
SYMPTOMS	Insomnia can be a symptom of GERD (Gastrointestinal Reflux Disease), asthma, heart arrhythmias, sleep apnea, menopause, depression, and/or thyroid disease. In addition insomnia can cause other illnesses and can be fatal for patients with diabetes or high blood pressure (hypertension).
TESTS	Usually your healthcare professional will order a sleep study.
TREATMENT	Historically the primary treatment has been limited to sleeping pills and cognitive behavioral therapy (i.e. a psychologist). However, if a cause can be determined (i.e. Sleep Apnea or Sleep Disordered Breathing etc.) it will be treated.

Note: Review sleep hygiene for better sleeping tips.

Interstitial Lung Disease (Parenchymal or Infiltrative Lung Disease)

DEFINITION	A group of several disorders that affect the small spaces of the lungs.
CAUSE	Inflammation of the interstitial spaces of the lungs; the air sacs (alveoli), the spaces around the small airways and blood vessels due to many different causes (i.e. organic dust, fumes, infection, radiation and many others).
SYMPTOMS	Symptoms will vary due to the cause however, shortness of breath (dyspnea), stiffening of lung tissue,
TESTS	Chest x-ray or CT scan to locate excess fluids, scarring (fibrosis), cysts (or honeycombing because it looks like a beehive), or lung shrinkage. Pulmonary Function Test (PFT) and exercise testing to test the lung's ability to function. Arterial Blood Gas (ABG) may show a low oxygen levels as well as a low carbon dioxide levels. Your healthcare provider may order a lung biopsy (see chapter 11) to test the lung tissue.
TREATMENT	Treat the cause and monitor the heart (the disease can be affected by the lung disease). Your healthcare may prescribe supplemental oxygen, bronchodilators (nebulizer), steroids (such as prednisone) to reduce inflammation, antibiotics if needed and mechanical ventilation in severe cases.

Ischemic Heart Disease (see Cardiovascular Disease)

DEFINITION A heart disease caused by reduced blood flow to the heart.

Ischemic Stroke (see Stroke)

Ketoacidosis (Diabetic), (also see Diabetes Mellitus (DM))

DEFINITION   Metabolic acidosis (increased acid levels not due to respiratory) is common to diabetes mellitus patients due to excess ketones (basically carbon).

Note: Breathing faster removes excess carbon dioxide, which increases pH (makes your blood more alkaline or less acidic).

Kyphoscoliosis (Combination of Kyphosis & Scoliosis), (also see Scoliosis)

DEFINITION   Kyphoscoliosis is a combination of Kyphosis (anterior to posterior or front to back) curvature and scoliosis (lateral or right to left) curvature of the spine (see diagram below).

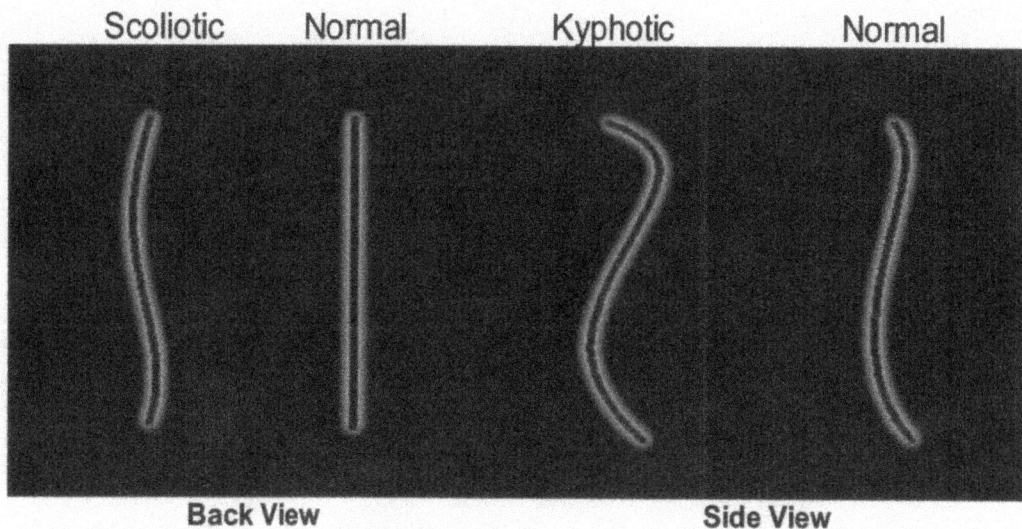

CAUSE   Congenital defect (birth defect), neuromuscular disorder (muscles pulling), TB of the bone (destruction of bone), or idiopathic (for no reason).

SYMPTOMS   Visible spinal and/or thoracic deformity causing, dyspnea (shortness of breath or difficulty breathing) on exertion, increased work of breathing, hypercapnia (elevated carbon dioxide level in the blood), hypoxemia (low blood oxygen level) and atelectasis (airless portion of lungs).

TESTS   Chest x-ray (see drawing above). Monitor for sufficient ventilation (Vital Capacity or the maximum amount of air you can breathe in the out) and Arterial Blood Gas (ABG) to determine hypercapnia (elevated carbon dioxide level in the blood) and hypoxemia (low blood oxygen level).

TREATMENT	Assisted breathing using BiPAP or mechanical ventilation (see chapter 6). Supplemental oxygen as required, pulmonary hygiene (movement of secretions) and treat any subsequent respiratory infections as needed.

Legionnaire's Disease (see Pneumonia)

DEFINITION	Pneumonia caused by the inhalation or ingestion of the bacterial organism Legionella Pneumophila or others.
CAUSE	A variety of organisms can be responsible for being febrile (high fever) and the respiratory infection that develops into pneumonia.
SYMPTOMS	Flu-like symptoms; fever, chills, headache, cough, fatigue, malaise (discomfort or uneasiness), and myalgia (muscle aches).
TESTS	Laboratory blood and sputum testing are used to confirm the presence and type of the organism. A chest x-ray reveals bilateral (in both lungs) patchy infiltrates (white patches) with possible consolidation (excess fluids in lung tissue). Breath sounds reveal rales/rhonchi (wet lungs), and wheezes (narrow airways).
TREATMENT	Treatment usually requires antibiotic therapy. In the worst case scenario the infection may progress causing respiratory failure. Usually only supplemental oxygen and pulmonary hygiene (moving of secretions) is necessary.

Lou Gehrig's Disease (see Amyotrophic Lateral Sclerosis (ALS))

Lung Abscess

DEFINITION	Lung abscesses have two categories Primary (starting within the lung) and Secondary (caused from outside the lung). The death (necrosis) of lung (pulmonary) tissue leaves cavities or pits which fill with fluids.
CAUSE	Most common in patients with periodontal disease, caused by aspiration (sucking bacteria or foreign materials (stomach contents) into the lungs). It may also be caused by pneumonia, bacteria, strep, fungus, tumors, septic emboli or parasites.
SYMPTOMS	Productive cough (purulent sputum or smelly and bad tasting mucus), shortness of breath (dyspnea), fatigue, fever, night sweats, coughing up blood (hemoptysis) and/or chest pain.
TESTS	Laboratory blood tests for inflammation markers (CRP or C - reactive protein), mucus or sputum cultures (growth, to identify infections) and chest x-ray to identify the area of infection. Breath sounds over the affected area are decreased with crackles (see chapter 8).
TREATMENT	Your healthcare provider will prescribe a broad-spectrum IV or oral antibiotics (kills or helps you fight many bacterial infections) until the x-ray shows the infection has cleared, and in sever cases surgical procedures for drainage or pulmonary resection or lobectomy may be required (see chapter 11). Most healthcare providers do not prescribe chest physiotherapy and postural drainage (see chapter 11) to remove the puss and mucus from the lungs due to the possibility of contamination of healthy lung tissue and the risk of obstructing small airways.

Lung Cancer/Adenoma/Neoplasm (see Bronchogenic Carcinoma & Tumors of the Lung)

Lung Malformations NEONATAL

DEFINITION   Congenital abnormalities (birth defects) effecting the lungs.

CAUSE   Lobar cysts (growth on the lungs), pulmonary sequestration (a nonfunctioning part of the lung that still has blood supplied to it) and lobar emphysema (not affecting the whole lung).

SYMPTOMS   Dyspnea (difficult breathing).

TREATMENT   Supportive care and surgery (see chapter 11).

Meconium Aspiration Syndrome (MAS) NEONATAL

DEFINITION   The material that collects in the intestines of a fetus and forms the first stool is expelled into the amniotic fluid and aspirated (enters the lungs).

CAUSE   The fetus actually inhales meconium, causing obstructions in the lungs, cysts or pneumothorax (air between the chest wall and lungs).

SYMPTOMS   Fetal tachycardia (rapid heartbeat) and the absent of fetal cardiac accelerations during labor (the baby is stressed during the birthing procedure and the heart rate should increase). After delivery, gasping respiration's, tachypnea (rapid breathing), grunting and retractions (using chest wall and neck muscles to breathe) should not be present.

APGAR SCORE			
	Points		
	0	**1**	**2**
Signs Checked			
Heart Rate	Absent	<100	>100
Respiratory Effort	Absent	Irregular & Slow	Strong Cry
Muscle Tone	Limp	Some Movement	Active
Irritability	No Response	Some Irritability	Vigorous
Color	Pale or Blue	Pink & Blue	Pink All Over

1 & 5 minutes after delivery

TESTS   Observation of meconium in mouth or nose. Chest x-ray shows patchy atelectasis (airless portion of the lungs) and areas of hyperinflation (over inflation). Arterial Blood Gas (ABG) of the umbilical artery indicates Acidosis (higher than normal acid level or decreased pH). An APGAR <5 (see APGAR above) is a medical emergency. Arterial Blood Gas (ABG) hypoxemia (low blood oxygen level) with Respiratory & Metabolic Acidosis (this is an uncompensated mixed acidosis, caused by not breathing well and the body building up acids due to muscle movement).

TREATMENT   Immediate suction oropharynx (back of the throat) at delivery, deliver blow by humidified oxygen (using an oxygen extension hose or corrugated tubing), insert ET tube and suction (2-4 times). If patient does not perk up, start CPAP (Continuous Positive Airway Pressure) to assist breathing or mechanical ventilation may be required.

Mesothelioma (Asbestosis)

DEFINITION	Fibrous pneumoconiosis (fibrous spots within the lung tissue).
CAUSE	Long-term exposure to and inhalation of asbestos dust.
SYMPTOMS	Dyspnea (shortness of breath or difficulty breathing) and a marked reduction in exercise tolerance. Cough and wheezing may occur in heavy smokers. Lung cancer may develop. Ultimately, respiratory failure will develop and will result in death.
TESTS	Chest X-ray will show small linear opacities (white spots) in the lower lung areas (lower lobes).
TREATMENT	Preventions is the best bet. Resulting cancers may be treated with surgery but surgery is ineffective in treating Mesothelioma. Supplemental oxygen and nebulizer (breathing) treatments will be ordered as required.

Myasthenia Gravis (Adult, Child & Neonate)

DEFINITION	A disorder of neuromuscular conduction that leads to muscle weakness of the skeletal muscles starting at the head and descending down through the body.
CAUSE	Autoimmune disorders are suspected causing a reduction in transmission of nerve impulses. Newborns can have a rare congenital form, which resolves in days to weeks.
SYMPTOMS	Usually involves the face, throat and ventilatory muscles and causes weakness and fatigue with any exertion. Early signs are similar to signs of a stroke (i.e. facial muscle weakness), ocular muscle weakness (droopy eyelids or double vision), and dysphasia (difficulty swallowing). Patients may progress into respiratory failure.
TESTS	Diagnostic testing will include the Tensilon Test (a short-acting anticholinesterase medication). If Tensilon improves the patient's condition temporarily, it is classified as a myasthenic crisis but, if the test worsens the patient's condition, it is classified as a cholinergic crisis.
TREATMENT	The patient will be monitored for ventilatory effort and pulmonary deterioration (using Vital Capacity, see chapter 11) to determine if mechanical ventilation support is required. Your healthcare providers will prescribe cholinesterase inhibitors and plasmapheresis (laboratory test for protein, removal of the autoanibodies and the replacement of the blood), corticosteroids, and immunosuppressive which may alter the disease course.

Myocardial Infarction (MI or Heart Attack)

DEFINITION	The lack of blood flow to the heart muscles usually causing permanent damage to the muscles due to the death of muscle tissue.
CAUSE	A complete or partial reduction of oxygenated blood flowing through the coronary arteries. This can be caused by fat, clots, proteins, air or calcium becoming stuck (most commonly in the coronary arteries)

	cutting off or reducing the blood flow to the heart muscles (see chapter 8 & 11 and drawing below).
SYMPTOMS	Each coronary artery supplies blood to different parts of the heart muscle (see drawing below) and symptoms differ in men and women. Because of the various blood supplies, symptoms may be different depending on where the blockage is located. During the heart attack symptoms will last 30 minutes or longer and will not be relieved with rest or medications (like angina, see angina). Symptoms may include chest pain, discomfort, pressure or heaviness (like an elephant is setting on you chest. The Pain may radiate to your back, jaw, throat or arm. You may also feel full or like you have indigestion or experience a choking sensation. Additionally you may experience nausea, vomiting, dizziness, sweating, esteem weakness, shortness of breath, rapid heart beat, irregular heartbeat and/or anxiety.
TESTS	Laboratory blood tests for a cardiac enzyme (released only during a heart attack) indicate heart damage. EKG (electrocardiogram, see chapter 8 & 11) will determine if your heart is working properly and will display a "Q" wave which indicates damage to heart cells (this change in the "Q" Wave may take hours to develop, see chapter 11). An echocardiography imaging test will determine how well the heart is pumping during and after a heart attack.
TREATMENT	Aspirin should be taken immediately, antiplatelet therapy to prevent blood clots, and thrombolytic therapy (clot busters are used to break up any blood clots that have formed. Your healthcare provider may use other medications to decrease the work of the heart, manage hypertension (high blood pressure), dilate blood vessels (vasodilators) and pain medications to manage pain. Cardiac catheterization may also be used to relieve ischemia (low blood flow to heart muscle) through angioplasty or stents (see chapter 11). In severe cases cardiac bypass surgery (see chapter 11) may be required to restore the blood flow through the coronary arteries.

Note: Hypertension (high blood pressure), type 2 diabetes, heart failure, stroke, arrhythmias (unusual heartbeat) and nocturnal complex arrhythmias (unusual heartbeat at night) have been associated with Sleep Disordered Breathing (SDB) also know as Sleep Apnea.

Note: The symptoms in men and women will vary. In men symptoms can be nausea, sweating, pain in the chest, jaw, arm or back depending on where the blockage or blockages are located. In women the symptoms include fatigue, depression and jaw pain (they do not normally experience chest pain).

Note: Patients with heart disease and erectile dysfunction (ED) may be at greater risk of having a heart attack. Talk to your healthcare provider about your risks and what you can do to help prevent a heart attack.

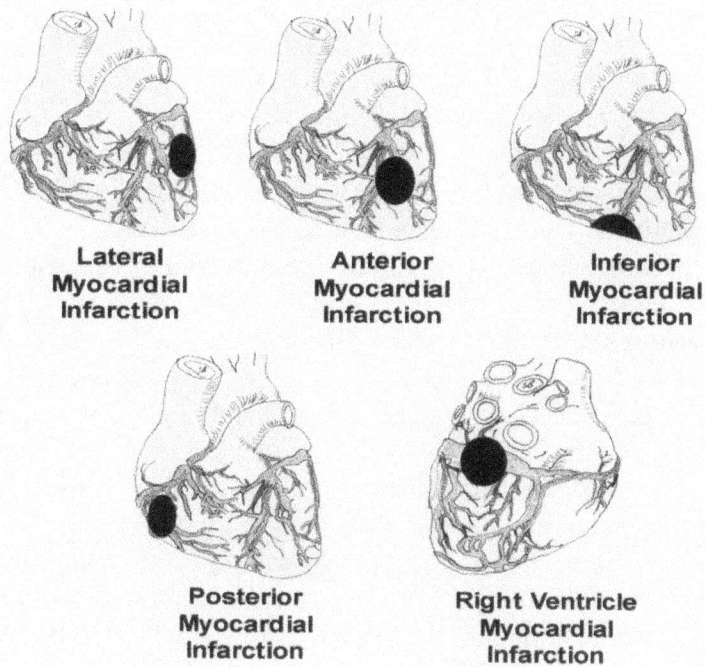

Blockages Causing Heart Attacks

Note: Great strides in stem cell research have identified their ability for rebuilding damaged heart muscle.

Neoplasm (see Bronchogenic Carcinoma)

Neuromuscular Disease (see Guillain-Barre Syndrome, Myasthenia Gravis, or Amyotrophic Lateral Sclerosis (ALS))

North-American Influenza (H1N1, H1N3, Pig Flu, Hog Flu, see Swine Flu)
Obstructive Lung Diseases (see specific disease)

DEFINITION   Diseases that reduce airflow through airways (breathing tube).

CAUSE   Asthma, Emphysema, Bronchitis, Chronic Obstructive Pulmonary
                    Disease (COPD), Mucus Plugging, Bronchiectasis, Bronchiolitis,
                    Allergic Bronchopulmonary Aspergillosis, and Cystic Fibrosis (in some
                    cases).

Note: New research has developed a test to detect genetic changes in the tissue of the trachea
of smokers in order to screen for early signs of cancer.

Occupational Asthma (see Asthma)

CAUSE   Work related asthma is divided into three groups. Occupational asthma,
                    work aggravated asthma and reactive airway disease. All of these may
                    adversely affect the health of workers.

148

Occupational Lung Diseases (see specific diagnosis)

DEFINITION   Lung disorders directly related to the environment at work (i.e. air contaminates, ionizing radiation etc.).

CAUSE   Inorganic Dust (see Silicosis and Coal Workers Pneumoconiosis), Organic Dust (see Occupational Asthma and Byssinosis) and/or Irritant Chemical/Gases (see Acute Exposure and/or Chronic Exposure).

Oxygen Toxicity (also see ARDS) ADULT - NEONATE

DEFINITION   A progressive respiratory failure that develops when high concentrations of oxygen (FiO2) >60% are used for a prolonged period.

CAUSE   Toxicity is based on the fact the body is not designed to process extremely high levels of oxygen for long periods. The toxicity can lead to lung damage and respiratory failure which results in a decreased oxygen tension (low oxygen level) within the blood. Additionally, in newborns blindness can result from a condition called ROP (Retinopathy of Prematurity or Retrolental Fibroplasia).

SYMPTOMS   The patient may exhibit cough, nausea, vomiting, substernal pain (chest pain), rapid breathing (tachypnea >20), may develop atelectasis (airless portions of the lung), decreased lung compliance (stiffer lungs), pulmonary edema (extra fluids in the lung tissue), pulmonary hemorrhage (bleeding in lungs), hypoxemia (low blood oxygen level), fibrosis (abnormal fiber tissue growth) and respiratory distress syndrome (see RDS).

TESTS   Arterial Blood Gas (ABG) will confirm the blood oxygen level. X-ray may reveal atelectasis (airless areas in the lungs).

TREATMENT   High levels of oxygen are seldom required over long periods. The use of CPAP (Continuous Positive Airway Pressure), BiPAP (same as CPAP but has separate inhalation and exhalation pressures) and/or PEEP (Positive End Exhalation Pressure) therapy (equipment discussed in chapter 6) is used to reduce oxygen requirements which saves money and much safer than using excessive levels of oxygen.

PAD & PAOD (see Peripheral Artery Disease)

Palpitations of the Heart (see Heart Palpitations)

Paracoccidioidomycosis (see Blastomycosis for North America)

DEFINITION   An infectious disease occurring only in South America, primarily involving skin, mucous membranes, lymph nodes and other internal organs.

CAUSE   Believed to be acquired by inhaling a fungus found in South & Central America where coffee-growers seem to be at the most risk.

SYMPTOMS   There are four clinical forms so symptoms range from sores under the skin on the face, around the nose and mouth, slowly expanding to ulcers with pinpoint yellowish-white areas, swelling of lymph nodes, to infection of internal organs.

| TESTS | Laboratory testing of the pus, biopsy tissue (tissue removed) and or cultures (growing the samples collected to determine what they are) are diagnostic (used to diagnose). |
| TREATMENT | Long-term medications are often required. |

Paradoxical Vocal Fold Motion (PVFM) see Vocal Cord Dysfunction (VCD)

Pectus Carinatum (Pigeon Breast)

| DEFINITION | A protruding chest abnormality of the sternum (center bone of the chest). |

Pectus Excavatum (Funnel Chest)

| DEFINITION | A concave (dip in) chest abnormality of the sternum. |

Pericardial Effusion (Cardiac Tamponade)

DEFINITION	The abnormal collection of fluids in area around the heart (normally 15 to 50ml or 1 to 3.3 tablespoons) causing pressure outside the heart (there are four types).
CAUSE	Bleeding (hemorrhaging) due to trauma, lupus, aneurysm (swollen blood vessels) or cancer, Congestive Heart Failure (CHF), tuberculosis. It can also be caused by Pericarditis, infections (bacteria and viral), Trichinosis (a parasite from undercooked pork), heart surgery, medications (some hair loss prevention medications), Hypothyroidism (reduced activity of the thyroid), Kidney Failure and others.
SYMPTOMS	Can be asymptomatic (without symptoms) to severe chest pain depending on the amount of fluid compressing the heart.
TESTS	A chest x-ray may show the shape of a flask also called water bottle heart.
TREATMENT	If treatment is required, your healthcare provider will treat the cause of the excess fluid (i.e. antibiotics or antiviral medications etc.). Your healthcare provider may prescribe anti-inflammatory medications and in more sever cases the fluid will be drained (by needle aspiration or catheter, called pericardiocentesis, see chapter 11) or in more severe cased surgery may be required.

Pericarditis

DEFINITION	The inflammation (fluid buildup) of the fibrous sac surrounding the heart (pericardium) and classified depending on the type of fluid causing the inflammation (exudate).
CAUSE	Can be caused by viral infection, bacterial infection, fungal infection, idiopathic (having no apparent cause) trauma (damage to the chest or heart), or heart attack, to name a few.
SYMPTOMS	The classic symptom is a friction rub (like pleurisy, discussed in chapter 8), chest pain (lessening when setting up and leaning forward) radiating to your jaw, back or left arm, difficult breathing (dyspnea), dry cough, fever, fatigue and or anxiety (nervousness). Anxiety may also make you experience impending doom (feeling like you are going to die)

| TESTS | EKG to look for changes (elevated ST segment and depressed PR segment, see chapter 11), swelling of ankles (edema), swollen neck veins (jugular distension), chest x-ray, and laboratory blood tests. |
| TREATMENT | Viral and idiopathic (of unknown origin) Pericarditis is treated with non-steroidal anti-inflammatory medications and bacterial Pericarditis will be treated with antibiotics and non-steroidal anti-inflammatory medications. More serious cases may require steroids and/or surgery. |

Peripheral Artery Disease (PAD) also within this category PAOD PVD

DEFINITION	PVD includes all dieses caused by the obstructions of large arteries in the arms and legs.
CAUSE	It results from trauma, atherosclerosis and inflammatory processes leading to embolism (clot blocking a vessel in the heart lung, brain etc.), thrombus (clot formation traveling through the body causing coagulation) or stenosis (narrowing of the vessel). As discussed in Atherosclerosis, things that place you at risk are; high cholesterol, overactive thyroid (hyperthyroidism), diabetes, obesity, smoking, high blood pressure (hypertension), aging, heredity, being male, sedentary lifestyle (not exercising), stress, sleep disorders, and some Sexually Transmitted Diseases (STD).
SYMPTOMS	Symptoms range from mild pain when walking (claudication), severe pain when walking, and pain while resting to tissue loss (gangrene) due to decreased blood flow. In addition, slow or non- healing sores (ulcers), blue or pale skin, cold extremities, and slowed hair or nail growth.
TESTS	If PVD is suspected an ankle brachial pressure index study will detect the reduction in blood pressure in the arteries supplying the legs. Doppler ultrasound and angiography (an x-ray using a contrast agent or fluoroscopy), and/or CT Scans (computerized tomography) may be used to locate the constriction.
TREATMENT	Prevention would be best, quit smoking, exercise, eat healthier, and create a healthier lifestyle. Angioplasty (balloon to reopen the constriction) or surgery to remove or bypass the constriction may be required and in severe cases amputations may be required. Medications to thin your blood, reduce cholesterol and breakup clots may be prescribed by your healthcare provider.

Peripheral Artery Occlusive Disease (see Peripheral Artery Disease

Persistent Pulmonary Hypertension of the Newborn (PPHN) NEONATAL

DEFINITION	A life threatening disorder, which reduces blood flow significantly from the lungs to the heart.
CAUSE	The newborn returns to fetal conditions in which right to left shunting (venous to arterial leak diluting the oxygen in the blood) takes place.
SYMPTOMS	Rapid changes in SpO2 (oxygen sampling finger probe) independent of changes in FiO2 (inspired oxygen), and severe hypoxemia (low blood oxygen levels). A significant shunt through the ductus arterious (baby

	heart/lung bypass) can be found by comparing two SpO2 monitors, one on the right arm and the other on either leg. Lung sounds may be clear.
TESTS	Chest x-ray may not be indicative or may indicate a congenital diaphragmatic hernia. EKG may help rule out congenital heart disease although occasionally cardiac catheterization (a long catheter is placed into the heart) may also be needed (see chapter 11).
TREATMENT	Hyperventilation (deep rapid breathing) with 100% oxygen to produce respiratory alkalosis (rapid breathing reduces the carbon dioxide and rises the pH) to reduce pulmonary vascular resistance. Vasodilatation (reducing blood pressure by making veins wider or larger) the use of alpha-adrenergic blockers may be tried with volume expanders (IV fluid). Care must be taken in the maintenance of fluids not to dilute electrolytes (sodium chloride, calcium etc), and glucose (sugar) or to overload the body with fluids.

Pertussis (Whooping Cough)

DEFINITION	Highly contagious bacterial infection lasting 6 weeks getting its name from the high pitched "WHOOP" sound upon inhalation following the cough.
CAUSE	Highly contagious contact or airborne respiratory infection requiring an incubation period of seven to ten days. This disease normally affects children and teens.
SYMPTOMS	Symptoms are nondescript flu like (low grade fever, sneezing runny nose etc.) and will include coughing fits stimulated by movement, talking, yawning, sneezing or eating followed by vomiting. The cough is unique, the cough is followed by difficulty breathing in which sounds like "WHOOP". Once you have heard it you will never forget the sound.
TESTS	Your healthcare provider will order laboratory culturing (growing) if nasal and/or mouth swabs.
TREATMENT	Prevention is the best bet by being vaccinated which will last for a few years requiring a booster. Antibiotics can shorten and/or lessen the severity of the disease and cough medicine or herbal treatments with vitamin C can decrease the severity of the cough.

Pickwickian Syndrome (Obesity Hypoventilation), (also see Sleep Apnea).

DEFINITION	Hypoventilation (shallow breathing) associated with obesity due to weight of the chest and excess tissue within the upper airways.
CAUSE	Obesity causing an upper airway obstruction during sleep and an excess massive chest tissue making it difficult to breathe properly.
SYMPTOMS	Snoring and apnea (stopping of breathing due to obstructions see Sleep Apnea) and/or air hunger (feeling like you cannot get enough air).
TESTS	Overnight sleep study without supplemental oxygen to confirm obstructive sleep apnea (stopping breathing due to airway obstruction, associated with snoring) and desaturations (decreased blood oxygen levels) due to apnea (stopping of breathing) episodes (see Sleep Apnea).
TREATMENT	Weight reduction (can correct the condition), CPAP Continuous Positive Airway Pressure) or BiPAP (same as CPAP but has separate inhalation

and exhalation pressures) to help maintain an open airway during sleep periods and supplemental oxygen may be required.

Plague (Bubonic, Septicemic, Pneumonic)

DEFINITION | The Bubonic Plague is a life threatening infection caused by Yersinia Pestis (Y-Pestis) carried by infected fleas. Septicemic Plague occurs when the plague bacteria infects your blood. Pneumonic Plague is a secondary infection of the lungs caused by the plague.

CAUSE | Pneumonic Plague is of high risk for use as a bio-weapon or bioterrorism affecting the lungs and spread from person to person through droplet contamination (cough).

SYMPTOMS | The symptoms will vary with the type of plague.

TESTS | Laboratory testing of blood and sputum.

TREATMENT | Antibiotics are effective in treating the plague and mechanical ventilation may be required to support breathing.

Pleural Effusion

DEFINITION | Any abnormal amount of pleural (lung) fluid in the pleural space (between the lungs and the chest wall). There are four different types of fluids that can accumulate in the pleural space are hydrothorax (clear or serous fluid), hemothorax (blood), chylothorax (chyle or triglycerides, i.e. oil), and/or pyothorax/empyema (pus).

CAUSE | There are many causes and diagnosis will be based on the type and quantity of fluid in the pleural space.

SYMPTOMS | Dyspnea (shortness of breath or difficulty breathing), tachypnea (rapid breathing >20), chest pain, cough, hemoptysis (bleeding from the lungs) and hypoxemia (low blood oxygen level). Breath sounds are decreased, with crackles over affected area. The increased the work of breathing (will wear the patient out) and decrease the lung's ability to intake oxygen and release carbon dioxide are critical factors.

TESTS | Chest x-ray may reveal the location of effusion with a Radiopaque mass (white area), obliteration of Costophrenic angle (the right and left side of the diaphragm can not be seen), mediastinal shift (lung tissue and trachea is shoved over) away from affected area and the possibility of atelectasis (airless portions of the lungs). Arterial Blood Gas (ABG) reveals hypoxemia (low blood oxygen level) and hypercapnia (elevated carbon dioxide level in the blood). Hand held ultrasound machines (works like a sonar using sound waves to locate the affected area) can greatly enhance the result of the thoracentesis treatment (see chapter 6 & 11).

TREATMENT | Thoracentesis (using a needle to draw of fluids), or chest tube for drainage placed between the chest wall and lung (see chapter 6 & 11) and the administration or supplemental oxygen as required.

Pleuritis (Pleurisy or Pleural Friction Rub)

DEFINITION	A disease causing inflammation of the pleura (skin between the chest wall and the lungs) normally due to lack of lubricating fluid around the lungs.
CAUSE	Causes a friction between the lungs and the chest wall (plural friction rub). The lack of lubrication makes breathing very painful. Pleurisy can be caused by virus, bacteria, neoplasm (abnormal growth), or autoimmune diseases (i.e. lupus where the body attacks itself).
SYMPTOMS	Dyspnea (shortness of breath or difficulty breathing) and considerable chest pain that can begin suddenly (usually inspiration is most painful) or with a cough. Breath sounds are similar to that of large ropes rubbing at a dock or the sound of two balloons being rubbed together. Pneumonia often develops, as well as inflammation and infection. Did we say it hurts a lot?
TESTS	Chest x-ray may reveal pleural thickening, pleural effusion (pleural fluid in the pleural space between the lungs and the chest wall), and/or atelectasis (airless portions of the lung).
TREATMENT	Non-steroidal anti-inflammatory medications (NSAID's), antibiotics (for infections), deep breathing with splinting (using a pillow held over the affected area to reduce the pain) followed by a cough to move secretions (pulmonary hygiene).

Pneumoconiosis

DEFINITION	Any disease of the lung caused by chronic inhalation of inorganic dusts, usually mineral dust of occupation or environmental origin.
CAUSE	Prolonged exposure with inhalation of certain dusts or chemical fumes causing lung disease such as asbestosis, silicosis, coal worker's disease (black lung) and others.

Pneumonia

DEFINITION	An inflammatory process of the lung parenchyma (essential or distinctive tissue of an organ), usually infectious in origin.
CAUSE	There are so many causes for this disease (i.e. bacteria, virus, fungal causative agents such as chemicals etc.). However, it all boils down to an inflammation of the alveoli (air sacs deep in the lungs), interstitial tissue, and bronchioles due to whatever. Overcoming the body's defenses, the cause of the trapping the agent in the lungs causing inflammation and infection.
SYMPTOMS	Bacterial infections are rapid (acute) in making you sick and the others develop more slowly. Symptoms will include high fever, productive cough (mucus production), shaking chills, dyspnea (shortness of breath or difficulty breathing), chest pain, inspiratory pain, fatigue, and many other flu-like symptoms. Breath sounds reveal crackles, rhonchi, and/or wheezing.
TESTS	Laboratory testing of sputum culture (growing the germs in the sputum) and blood samples can identify the type of infection. Chest x-ray will reveal infiltrates (diffuse white areas) in the infected area.

TREATMENT   A vaccine for patients >65 years old for prevention is recommended. Your healthcare provider may prescribe bronchodilators (nebulizers or MDI's), medication for fever reduction, corticosteroid, anti-inflammatory agents, antibiotics (for bacterial infections), pulmonary hygiene (moving secretions), deep breathing exercises (incentive spirometer) and supplemental oxygen if required.

Pneumonitis

DEFINITION   Inflammation or infection in the lungs (see pneumonia).

Pneumothorax

DEFINITION   The presence of air in the intra-pleural space (between the lungs and chest wall) is usually caused by trauma but can be spontaneous (just happens).

CAUSE   This can be an acute medical emergency depending on how much air is present. The air within the pleural space cannot escape and can make breathing extremely difficult due to restricting the lung's movements.

SYMPTOMS   The immediate (acute) onset of dyspnea (shortness of breath or difficulty breathing), pain, tachypnea (rapid breathing >20), increased work of breathing, tachycardia (rapid heart rate >100), and hypoxemia (low blood oxygen). This condition can result in respiratory failure. Breath sounds are diminished or even absent over affected area.

TESTS   Chest x-ray reveals hyperlucency (white spaces) and absent vascular markings (veins and lymph system) within the area where the air is located, the trachea may be deviated away from affected side, and/or atelectasis (airless portion of the lungs) may be revealed.

TREATMENT   Small pneumothorax can resolve without treatment although, thoracentesis (needle aspiration of the air, see chapter 11), chest tube (see chapter 6 & 11), and supplemental oxygen may be required.

Popcorn Lung (see Fixed Obstructive Lung Disease)

DEFINITION   Rare yet potentially deadly type of COPD known as Fixed Obstructive Lung Disease.

CAUSE   Occupational exposure without respiratory protection to chemicals used in food flavorings.

SYMPTOMS   Nonproductive cough (no sputum), dyspnea (shortness of breath or difficulty breathing with exertion (exercise) and/or wheezing. Some patients may experience night sweats, weight loss, eye/nose irritation and/or fever.

TESTS   Pulmonary Function Test (see chapter 6 & 11) indicates obstructions, hyperinflation (too much air due to air trapping), and FEV1 (rapid exhaling) test lower than normal. Chest x-ray may show hyperinflation (too much air due to air trapping) or may look normal. CT scan may reveal air trapping, thickened airway walls and/or general haziness. Lung biopsies (lung tissue samples) may constrictive Bronchiolitis Obliterans (narrowing or complete obstructions in the small airways deep in the lungs).

155

TREATMENT	Prevention is the best bet. Medication may be prescribed to relive the symptoms (see COPD).

Psittacosis (Chlamydia Psittaci Infection, Parrot Fever or Ornithosis)

DEFINITION	Intracellular parasite infection caused by the inhalation of Chlamydia Psittaci, which is a gram-negative bacterium, found in bird droppings.
CAUSE	Can result in a long lasting illness, which can progress into pneumonia.
SYMPTOMS	Can be asymptomatic (no symptoms) or flu-like with patients suffering from headache, cough, dyspnea (shortness of breath or difficulty breathing), hypoxemia (low blood oxygen), tachypnea (rapid breathing >20), sore throat, nausea, vomiting, epistaxis (nosebleed), constipation, anorexia, myalgia (tenderness/pain in muscles or aches/pains), and possibly severe pneumonia. Respiratory failure is possible in extreme cases.
TESTS	Laboratory testing via blood culture, positive for Chlamydia will confirm the diagnosis.
TREATMENT	Antibiotics are required (i.e. tetracycline, erythromycin, or penicillin etc.).

Pulmonary Abscess

DEFINITION	A collection of purulent material (pus) in the lung which causes inflammation and tissue necrosis (death of tissue) due to aspiration (sucking in to lungs) of infectious organisms.
TREATMENT	Your healthcare provider will prescribe antibiotics and pulmonary hygiene (moving secretions).

Pulmonary Air-Block Syndrome (Pulmonary Interstitial Emphysema) Neonatal

DEFINITION	Air (air-leaks) outside of the normal pulmonary air spaces.
CAUSE	Probably the result of a large negative intrathoracic (within the chest) force while taking the first breaths of air.
SYMPTOMS	Many infants are asymptomatic (no signs) or have only tachypnea (rapid breathing).
TESTS	ABG (Arterial Blood Gas) may indicate acidosis (lower pH or an increased acid level), hypercapnia (elevated carbon dioxide level), and hypoxemia (low blood oxygen level). Chest x-ray reveals cystic or linear lucencies (white spots).
TREATMENT	The course is highly variable and may resolve over 1-2 days or persist for months. Some infants develop Bronchopulmonary Dysplasia (see BPD). Mechanical ventilation (at the lowest possible inspiratory pressure) in order for the lung/lungs to heal. Supplemental oxygen (at the lowest level possible) is usually required as well as other supportive care.

Pulmonary Alveolar Proteinosis

DEFINITION	A rare disease which fills the alveoli with proteins and phospholipids.

CAUSE	Unknown.
SYMPTOMS	Can begin as asymptomatic (without symptoms) or progress to limiting your activities due to dyspnea (shortness of breath).
TESTS	Chest x-ray may show a butterfly shape within the lungs, Pulmonary Function Test (see chapter 11) will indicate a reduce lung function and a lung biopsy (see chapter 11) will be ordered to test the lung tissue.
TREATMENT	Some patients do not require treatment. In more severe cases a lung lavage (basically washing or rinsing out the lungs) can resolve the condition in as little as one treatment or treatments may be required every six months over a period of years.

Pulmonary Edema

DEFINITION	A condition in which excessive amounts of plasma (fluid) enter the pulmonary interstitial (lung tissue) and alveoli (air exchange sacks deep in the lungs).
CAUSE	Can be caused by congestive heart failure (CHF), hypervolemia (excess blood in the blood stream) which causes hypertension (high blood pressure), high altitude (see note below), pneumonia, embolism, oxygen toxicity, smoke inhalation, sepsis (blood infection), renal (kidney) failure, heart disease, myocardial infarction (heart attack), ARDS (see ARDS), chemical inhalation or near-drowning (water in the lungs).
SYMPTOMS	Dyspnea (shortness of breath or difficulty breathing), hypoxemia (low blood oxygen), cyanosis (a blue tint to skin, lips or fingernails), productive cough (pink frothy secretions), hyperventilation (deep breathing), tachycardia (rapid heart rate >100), orthopnea (difficulty breathing if not setting or standing), and anxiety. The patients breath sounds reveal rales (wet lungs).
TESTS	The chest x-ray shown prominent vascular markings (veins look swollen), diffuse fluffy infiltrates (cloud looking white patches) in a butterfly pattern, and/or cardiomegaly (enlarged heart, the heart is normally about the size of your fist).
TREATMENT	Your health care provider will prescribe diuretics (water pills help remove excess water), digitalis (heart medication), morphine (pain and blood pressure reducer), and high-concentration oxygen until the condition reverses. Severe cases may require support by mechanical ventilation with PEEP (Positive End Expiratory Pressure) therapy which is designed to keep the airways open and help remove the excess fluids from the lung tissue (due to the artificial air pressure in the lungs (see chapter 6)) as required.

Note: High Altitude Pulmonary Edema (HAPE) is an acute (rapid onset) life threatening illness which develops into HACE (High Altitude Cerebral Edema or swelling of the brain) which is deadly. If you plan a trip to the Rockies or other high altitude vacation spot, include your healthcare provider in the planning. Sulfa medication can reduce your changes of developing HAPE. Additionally, acclimating to altitude (spend a day or two at 5280 feet before going up to the mountains) will reduce your chances of getting sick. This is good advice no matter what your age.

Pulmonary Embolism (PE), Thromboembolism

DEFINITION   The sudden partial to complete blockage of pulmonary artery blood flow.

CAUSE   Usually caused by a blood clot (thrombus) formed by a Deep Vein Thrombosis (DVT or clot from deep in your legs), fat, air or tumor (very rare).

SYMPTOMS   Dyspnea (shortness of breath or difficulty breathing), hyperventilation (deep breathing), tachycardia (rapid heart rate >100), chest pain, anxiety, arrhythmias (abnormal or irregular heartbeats), hypoxemia (low blood oxygen level), cyanosis (a blue tint to skin, lips or fingernails), cough, hemoptysis (bleeding from the lungs) and atelectasis (airless portion of the lungs) are common. Breath sounds reveal rales (wet lungs), rhonchi, and/or wheezes. This condition can cause pulmonary necrosis (lung tissue death), Cor Pulmonale (Right Ventricular enlargement secondary to a malfunction of the lungs), heart failure, and/or even respiratory failure.

TESTS   Diagnostic testing using a CT scan, ultrasound (uses sound waves like a sonar), ventilation perfusion scan (V/Q scan checks flows), and pulmonary angiography's (finds blockages and measure blood pressures, see chapter 8 & 11). Laboratory test the "D" Dimer (blood clot test) checks for the markers (a substance released when a blood clot breaks up).

TREATMENT   Prevention is the best bet. Patient education for using compression elastic stockings, leg exercises, and increasing your activity level are key. Your healthcare provider may prescribe oxygen, anticoagulants (blood thinners) and thrombolytic (clot busters) therapy. If all ease fails, a clot filter can be placed in a vein to trap clots. Ventilatory support using mechanical ventilation may be ordered if respiratory failure develops.

Note: Any long period setting/laying without exercise, (i.e. trains, planes, automobiles etc.) can be the cause. It is important to stand walk around or just exercise you legs every couple of hours to prevent clots from forming. During international flights the airline will show you an exercise video and encourage you to participate in the exercise program in order to help you prevent clots form forming. Prevention is easy and clots can kill…

Pulmonary Hypertension (also called Pulmonary Artery Hypertension)

DEFINITION   A very uncommon obliterative (destroys tissue) disease of unknown cause involving medium and small pulmonary arteries and terminating in right ventricular failure or fatal syncope (passing out resulting in death).

CAUSE   The blood pressure within the pulmonary arteries, veins and/or capillaries is high due to disease or disorders, which change the blood flow of lungs and heart. Due to this high backpressure, pulmonary embolism (blockage of the pulmonary artery by foreign matter), heart failure or cardiogenic shock (damage to the heart reduces blood blow provide from the heart) resulting in COPD, ARDS or pulmonary fibrosis.

SYMPTOMS	Shortness of breath (dyspnea) on exertion, light-headedness, weakness, chest pain, and peripheral edema (retention of water or swelling of the ankles or feet) often are present.
TESTS	A pulmonary artery catheter is used to measure blood pressures.
TREATMENT	Usually diuretics (to control water retention), anticoagulants (prevent clots), vasodilators (make veins larger to reduce blood pressure) and oxygen are ordered. The cause must be identified and treated by your doctor.

Pulmonary Vascular Disease (see Pulmonary Hypertension and Venous Thromboembolic Disease)

PVD (see Peripheral Artery Disease)

Respiratory Distress Syndrome (RDS) NEONATAL

DEFINITION	A disorder primarily of premature infants, developing respiratory distress.
CAUSE	A surfactant (lung lubricant) deficiency causes the alveoli (air exchange sacks deep in the lungs, see chapter 7) to collapse.
SYMPTOMS	Tachypnea (rapid breathing), grunting retractions (chest muscles suck in), nasal flaring (nose opens wide, like a rabbit's nose), labored breathing (due to stiffer lungs or less compliance), systemic hypotension (low blood pressure), hypothermia (low body temperature), and poor perfusion (the exchange of oxygen in the cells).
TESTS	Auscultation (breath sounds) reveals inspiratory crackles/rales (wet lungs). Chest x-ray shows diffuse hazy areas, atelectasis (airless portion of the lungs) worsening over time. Arterial Blood Gas (ABG) will show hypoxemia (low blood oxygen level) and hypercapnia (elevated carbon dioxide level in the blood).
TREATMENT	Monitor closely, transcutaneous monitors (continuously monitors oxygen and carbon dioxide through the skin without drawing blood) and provide supplemental oxygen as required. Mild cases do well with supplemental oxygen via an oxygen hood or Oxyhood (looks like a half of a fish bowl over their face) however, more sever cases will require Continuous Positive Airway Pressure (CPAP) or mechanical ventilation to keep airways open (see chapter 6). Surfactant is added to the lungs via an endotracheal tube (see chapter 6) to help restore the lung's lubrication balance.

Respiratory Failure

DEFINITION	Respiratory failures are divided into two types, the inability to maintain adequate oxygen supply to tissues/organs (hypoxemic respiratory failure) and the inability to remove carbon dioxide from tissues/organs (hypercapnic respiratory failure).
CAUSE	Both can be an acute (rapid onset) medical emergency or chronic condition monitored by your healthcare provider. Disorders of the lungs or medical complications are the primary factors causing this condition.

SYMPTOMS	Symptoms vary between asymptomatic (no symptoms) to death depending on degree of hypoxemia (low blood oxygen level) and hypercapnia (elevated blood carbon dioxide levels). Other symptoms include; dyspnea (shortness of breath or difficulty breathing), anxiety, and respiratory distress to unconsciousness and death.
TESTS	Arterial Blood Gas (ABG) is analyzed to determine severity.
TREATMENT	Supplemental oxygen and BiPAP (continuous positive airway pressure with inhalation pressure as well as a lower exhale pressure) to full ventilatory support may be required (see chapter 6 for equipment).

Respiratory Syncytial Virus (RSV) NEONATAL/ PEDIATRIC/ADULT

DEFINITION	A common virus causing mild to sever respiratory infections. Nearly all infants have had RSV by age two.
CAUSE	Usually during the fall, RSV causes pneumonia, bronchiolitis (inflammation of the small airways of the lungs) and croup (the one with the barking cough. Usually it is a mild respiratory illness in adults and older children.
SYMPTOMS	Symptoms will drastically vary, differing with age but may include; inspiratory wheezing (narrowing airways), nasal congestion, nasal flaring (nares or nostrils open and close), cough, tachypnea (rapid breathing) labored breathing, dyspnea (shortness of breath or difficulty breathing), cyanosis (a blue tint to skin, lips or fingernails), fever, possibly a croupy cough, and increased airway resistance progressing to respiratory failure due to fatigue in severe cases.
TESTS	Chest x-ray pneumonia or Bronchiolitis and/or areas of hyperinflation (over inflated). Arterial Blood Gas (ABG) shows hypoxemia (low blood oxygen level). The laboratory can complete serologic virus isolation test on blood or sputum, which may help to identify the virus.
TREATMENT	Antibiotics will not cure a virus. The antiviral medication Ribavirin has been proven effective in treating influenza A or B and RSV. Mild cases may be resolved using nebulizers, humidification and oral decongestants at home. More severe cases require a croup tent (the old moisturized oxygen tent) normally <48 hours, supplemental oxygen as required and antibiotics to control secondary bacterial infections.

Note: Respiratory Syncytial Virus (RSV) causes about 10% of colds in children.

Restrictive Disease (see condition specific information)

DEFINITION	Diseases that restrict the lungs ability to expand and contract.
CAUSE	Asbestosis, Fibrosis, Acute Respiratory Distress Syndrome, Hypersensitivity Pneumonitis, Pneumonias, Sarcoidosis, Pulmonary Alveolar Proteinosis, Neuromuscular Diseases, and Infant Respiratory Distress Syndrome.

Note: New research has developed a test to detect genetic changes in the tissue of the trachea of smokers in order to screen for early signs of cancer.

Rhinovirus (Common Cold) causes about 50% of all colds

DEFINITION   The common cold consists of over 200 possible viruses which cause upper respiratory tract (lungs, trachea (or breathing tube) and throat).

CAUSE   Wet hair and being cold does not cause colds. You get cold from contact with the virus (i.e. mouth and nose) and you are at greater risk when you have allergies, fatigued (tired) or have emotional distress. Once the virus has attached itself to the lining of you mouth, nose or lungs the virus begins to replicate itself.

SYMPTOMS   Sore throat, sneezing, nasal congestion, runny nose, mucus drainage, watery eyes. More severe symptoms such as high fever (febrile) $\geq$100 degrees, and muscle aches might indicate the flu (influenza).

TESTS   Your healthcare provider may order laboratory tests for sputum (mucus from a deep cough) or blood tests to identify the infection. However these tests take 24-72 hours for results.

TREATMENT   Your healthcare provider will most likely diagnose your illness based what illness has been common for your area. He/she may also order steroids and respiratory medications to improve you breathing. The best advice is not to get sick, wash your hands frequently, stay away from sick people, cough/sneeze into your elbow and keep your fingers out of your mouth and nose.

Note: The Center for Disease Control (CDC) states that older adults average less than 1 cold per year, adults average 3 colds per year, and children may have as many as 12 colds per year. Much of this depends on jobs, contact and where you go.

RSV (see Respiratory Syncytial Virus)

Sarcoidosis

DEFINITION   A chronic disorder of unknown origin characterized by the formation of tubercles (small nodules or growths) of necrotizing (killing) epithelia tissue (skin or mucus membranes).

CAUSE   Idiopathic (no cause identified). Possible causes could include a hypersensitive (over sensitive) response to environmental factors, genetic predisposition (inherited), or extreme immune response (body fighting itself) while fighting an infection. Disease characterized by granulomas (mass of cells), and inflammation or edema (swelling of tissue due to retention of fluids) in the lungs, lymph nodes, liver, eyes, skin, and many other tissues or organs.

SYMPTOMS   Often asymptomatic (no symptoms) but can manifest as cough, fever, malaise (fatigue), dyspnea (shortness of breath or difficulty breathing), hypoxemia (low blood oxygen level), swelling of the liver, enlargement of the lymph glands, spleen, or skin rash/lesions. Some patients have developed a pneumothorax (air between the chest wall and lungs), pulmonary hypertension (high pressure between the lungs and heart), Cor Pulmonale (right heart failure) or pulmonary fibrosis (abnormal growth of tissue). Rarely do patients develop complete organ failure.

TESTS   Tests are be based on symptoms, location and severity.

TREATMENT   Steroid therapy will reduce swelling and help to manage symptoms. Additionally, patient education in pulmonary hygiene (moving secretions) is of critical importance. Severe cases may prompt your healthcare provider to recommend a transplant of the affected organ/s.

Note: Sarcoidosis can be present in various organs or tissues with symptoms dependant on its location.

SARS (see Severe Acute Respiratory Syndrome)

Scoliosis (Including Kyphosis), (see Kyphoscoliosis for a diagram of spine appearance)

DEFINITION   Scoliosis is abnormal lateral (side-to-side) curvature of the spine (see figure under Kyphoscoliosis). Kyphosis is an abnormal anterior to posterior (front to back) curvature of the spine. Kyphoscoliosis is much more severe defect because it is the combination of kyphosis and scoliosis (see Kyphoscoliosis for a diagram of spine appearance). These spinal/thoracic deformities are usually restrictive in respect to ventilation and can dramatically increase the work of breathing.

CAUSE   Both can be due to congenital defects or Tuberculosis (TB), malignancy, or compression fractures.

SYMPTOMS   A visible spinal and/or thoracic deformity causing compression of the thorax resulting in, dyspnea (shortness of breath or difficulty breathing) with exertion, increased work of breathing, hypercapnia (elevated blood carbon dioxide level), hypoxemia (low blood oxygen), atelectasis (airless portion of the lungs)

TESTS   Visual observation, chest x-ray (see Kyphoscoliosis drawing) and Pulmonary Function Test (PFT) will show a restrictive disease.

TREATMENT   Monitor the patient's for sufficient ventilation and given supportive care as required. Patient education in pulmonary hygiene (secretion movement) and proper care at home is essential.

Severe Acute Respiratory Syndrome (SARS)

DEFINITION   Acute (rapid onset) upper respiratory illness, first appeared in Asia.

CAUSE   The coronavirus (under a microscope resembles the corona of the sun during an eclipse) is believed to have developed from eating infected animals in Asia. Infection may develop after sharing food/drink, personal contact (hugging & kissing), hand to mouth or nose transfer (touching infected surfaces) or close proximity to an infected person (less than 3 feet).

SYMPTOMS   Cold-like symptoms (developing over 3 to 10 days); fever higher than 100.4F, nonproductive cough (dry cough), dyspnea (shortness of breath or difficulty breathing), headache, muscle aches, sore throat, fatigue and/or diarrhea. The elderly may not develop a fever but, may have a feeling of malaise (uneasy or unwell feelings) and loss of appetite.

TESTS   Identification of travel to an infected area along with a fever higher than 100.4F. Chest x-ray may be done if dyspnea (shortness of breath or difficulty breathing) is present or to rule out other causes. Laboratory testing may include blood sample, sputum sample and/or nasal swab to

	detect the bacteria or virus. Additionally antibodies (fight the SARS infection) may be detected in your blood.
TREATMENT	Prevention is the best bet. Wash your hands frequently and do not come into contact with infected people. No medication is available to cure SARS however, antiviral medication (ribavirin) along with corticosteroids may help by increasing the amount of oxygen in the blood.

Silicosis

DEFINITION	Respiratory disease caused by silica (sand or what glass is made from) dust inhalation exposure.
CAUSE	Silica is a crystal found in beach sand, rock beds, mines, stone processing and desert regions. Exposure to a large amount quickly or small amount over a long period may result in inflammation, interstitial fibrosis (abnormal fiber growth in the lung tissue), and severe scarring of lung tissue.
SYMPTOMS	Symptoms will vary depending upon the exposure amount, time and the effect on the body. Patient will develop dyspnea (shortness of breath or difficulty breathing), a chronic cough, pulmonary edema (extra water in the lung tissue), progressive pulmonary fibrosis, and destruction of normal lung tissue.
TESTS	A chest x-ray will reveal a radiolucent (white areas) appearance in the affected area. Arterial Blood Gas (ABG) results will vary depending on how well the lungs are functioning. Breath sounds are very diminished over the affected area.
TREATMENT	Prevention is your best bet, no known cure exists. Your healthcare provider will prescribe Pulmonary Rehabilitation, Patient Education and supportive care.

Sleep Apnea (Obstructive Sleep Apnea (OSA), Central Sleep Apnea (CSA)) also known as Sleep Disordered Breathing (SDB)

DEFINITION	Sleep apnea is repeated episodes of complete cessation (stopping) of airflow for 10 seconds or longer. Sleep Apnea is divided into three groups. (1) Obstructive Sleep Apnea (OSA) which is due to upper airway closure resulting in snoring, (2) Central Sleep Apnea (CSA), a rare condition which is due to problems with the central nervous system (you just stop breathing) and (3) Mixed Apnea which is a combination of OSA and CSA.
CAUSE	Apnea (not breathing for 10 seconds or more) will cause a severe desaturations (decrease of your blood oxygen level) also called hypoxia (low blood oxygen level). Obstructive Sleep apnea is the most common type and usually is caused by obstruction of the upper airway during sleep due to obesity, narrowed airway, enlarged tonsils, enlarged adenoids and/or sedation (alcohol, etc.). Central Sleep Apnea (very rare) is a neurological disorder caused by brain stem, brain tumor, Ondine's curse, or idiopathic causes (no known cause) which results in central hypoventilation (shallow or slow breathing) where the brain fails to signal the body to breathe. Mixed sleep apnea (even rarer) is a

	combination of obstructive and central sleep apnea disorders. Additionally, children who are obese are more likely to develop sleep apnea.
SYMPTOMS	The best indication you have sleep apnea is that your spouse (or significant other) wakes you us and says "roll over", "stop snoring and breath", "you scared me" or "one of us is going to the couch". Sometimes you wake yourself up when gasping for air or making a lot of noise (you may wake often and not really know why). Additionally, you may have trouble waking in the morning, trouble staying awake during the day, experiencing fatigue, irritability, difficulty concentrating, confusion, headaches, and possibly hallucinations. These are dangerous symptoms and you need to discuss them with your healthcare provider in order to keep you and your family safe. Hypertension, type 2 diabetes, heart failure, stroke, arrhythmias and nocturnal complex arrhythmias have been associated with Sleep Apnea also know as Sleep Disordered Breathing(SDB). Additionally, Sleep Apnea, frequent waking or not sleeping may cause a common form of Dementia due to protein formation.

Note: Men and women who suffer from sleep apnea can double their risk for stroke, high blood pressure and heart disease. In addition sleep disorders can lead to severe headaches.

TESTS	An Overnight Sleep Study (Polysomnography) which monitors pulse ox (blood oxygen level), Heart Rate and Rhythm (EKG), brain waves (EEG), eye movement, chest movement, abdomen movement (breathing, coughing etc.), airflow sensor (air movement from mouth and nose), leg movements, and audio and video (sleepwalking, talking, asthma, etc.). Sleep apnea is classified as severe if you experience more than fifteen apnea episodes each hour. Other testing may be required, such as Multi-Sleep Lab Test (MSLT), which is used for diagnosing other sleep disorders.
TREATMENT	Obstructive Sleep Apnea (OSA) may be treated through weight loss, avoiding alcohol, avoiding cigarettes, avoiding use of sedatives, splinting of nose (opening the nose wider) and/or using CPAP/BiPAP (both are Continuous Positive Airway Pressure & BiPAP uses different inhalation and exhalation pressures (see chapter 6)) in order to maintain an open airway. Surgery is an alternative to reduce excess airway tissue, remove tonsils, and/or correct airway problems (consider weight loss or stop drinking first). Central Sleep Apnea (CSA) is usually treated with antidepressants, supplemental oxygen and/or CPAP/BiPAP.

Note: Snoring is a wakeup call; it is a warning to you that something is wrong. It is believed that sleep apnea can result in congestive heart failure (CHF), myocardial infarction (heart attack), heart failure, respiratory failure and has been linked to the worsening of diabetes. If it is coupled with COPD it is know as "Overlap Syndrome (see COPD).

Sleep Disordered Breathing (SDB) see Sleep Apnea

Note: Sleep disorders can lead to severe headaches.

Smoke Inhalation

DEFINITION   The inhalation of smoke into the lungs.

CAUSE   Smoke, caustic fumes from fire cause pulmonary (lung) damage.

SYMPTOMS   Symptoms and damage depend on the temperature, amount, length of time and type of smoke inhaled. Singed nose hair indicates high temperatures and may result in dyspnea (shortness of breath or difficulty breathing), hypoxemia (low blood oxygen levels) and damage to lung tissue, which may result in respiratory failure.

TESTS   Laboratory blood tests for carboxyhemoglobin (carbon monoxide >5%), Arterial Blood Gas (ABG) for oxygen and carbon dioxide level. Chest x-ray may reveal damage to lung tissue or fluid buildup (edema) within the lung and trachea. Observation of singed areas in and around the mouth and nose will help to determine how hot the smoke was when inhaled.

TREATMENT   Hyperbaric Oxygen (HBO) also called decompression chamber (therapy or oxygen supplied at 100%). Your healthcare provider may prescribe steroids to reduce swelling of lung and mucosal tissue. In severe cases, mechanical ventilation for respiratory failure may be required, (see chapter 6).

Smallpox (Variola Virus Minor and Major)

DEFINITION   A contagious, disfiguring and deadly disease (death rate is about 50 percent).

CAUSE   The naturally occurring or bio-constructed (man made) variola virus infection is spread when it is inhaled, swallowed or touched having an incubation period of 7 to 14 days.

SYM PTOMS   During the incubation period (7-14 days) you will have no symptoms and are not infectious. Then sudden fever (the point you become infectious), discomfort (malaise), severe fatigue (prostration), severe back pain, headache, nausea, vomiting and/or diarrhea. Within a few days of symptom onset the rash will appear, first as flat red spots or pinpoint spots, and then small blisters with clear fluids, followed by blisters filled with pus (pustules). These blisters will form head to toe, front to back, inside and outside of your body leaving pitting in the tissues of the survivors.

TESTS   Diagnoses is made by identifying the blisters.

TREATMENT   There is no treatment (except supportive care like antibiotics to treat other infections) or cure, prevention through vaccination is your best bet. A medication called cidofovir has shown promise in laboratory studies. You should consult your healthcare provider about vaccines if an outbreak should occur anywhere in the world (you have about 4 days to be vaccinated if exposed). The Smallpox vaccine contains live virus and can cause serious harm to your body. These vaccinations are most effective for 3 to 5 years and known to provide partial immunity for many more years.

Note: Smallpox was eradicated (killed off) worldwide in 1980 except for samples in two laboratories, one in Siberia Russia (tons of viruses) and one in the United States (medical samples). Why, you ask, do we mention Smallpox at all? Though it is believed the risk of using Smallpox for terrorism (biological warfare agent) is low, if used it would be devastating to the world (about 33 to 50 percent will die and many survivors will be blind).

Stable Coronary Artery Disease

DEFINITION Previous heart attack, bypass surgery, angioplasty, abnormal stress test or a narrow coronary artery.

Stroke

DEFINITION There are many types of strokes but, they all deprive brain tissue of oxygen and nutrients when blood supply to part of the brain is interrupted or reduced. This interruption begins damaging cells within 3 minutes.

CAUSE Major risk factors are smoking, high cholesterol, high blood pressure or head trauma. Ischemic Stroke (about 75% to 80%) are caused by blood clots (thrombus) or plaque blocking arterial flow the brain and an Embolic Stroke is when the blockage comes from somewhere else (like your heart) and lodge in an artery of the brain. A Hemorrhagic Stroke is a leaking or ruptured vessel within the brain (Intracerebral Hemorrhage or a Subarachnoid Hemorrhage. Strokes have been associated with Sleep Disordered Breathing (SDB) also known as Sleep Apnea.

SYMPTOMS Experiencing paralysis or numbness on one side of the body, trouble speaking, trouble walking, disturbances in sight, headache (you may see a bolt of light), vomiting, fainting, and/or dizziness. In most cases there is no warning and may last for minutes or longer.

TESTS Physical exam and medical history, Carotid Ultrasound, Arteriography, CT of neck and brain, MRI of neck and brain and/or Echocardiography, (see chapter 11 for test explanations). Laboratory blood tests for enzymes occurring during and after a stroke or blood clot (d Dimmer).

TREATMENT Always get help (call 911) immediately. Emergency medications (for strokes that are not active bleeds such a Hemorrhagic Stroke) like aspirin; Coumadin, Heparin and Tissue Plasminogen Activator (TPA) are all blood thinners (anticoagulants). In severe cases and in Hemorrhagic Stroke surgery may be required to correct the problem. Medications to control cholesterol and hypertension, reduction of water weight (diuretics), anticoagulants (to reduce clotting) and/or diabetes may be ordered. Lifestyle changes will be required for weight control, quit smoking/drinking, exercise routine, must manage stress, and if illicit drugs are used they must be stopped.

Sudden Cardiac Arrest (see Cardiac Arrest)

Sudden Infant Death Syndrome (SIDS) NEONATAL/ PEDIATRIC

DEFINITION SIDS is the leading cause of deaths of infants <1 year old (1 to 3 months are most at risk), about 7,000 per year or 40% of infant deaths.

CAUSE   Autopsy have shown that these unexpected deaths while sleeping have had repeated episodes of hypoxia (not enough oxygen available for the body) or ischemia (localized reduction in blood flow to an organ). Apnea of prematurity (lack of lung development of the newborn) is not a predisposing factor and no evidence has been found to suggest immaturity of the respiratory centers is related. Infants sleeping in the prone position (on their stomach) have been strongly associated with SIDS and are not recommended. Although, we just do not really know.

SYMPTOMS   Identification of patient history (Hx) may be valuable to determining risks. Higher risk exists in families with two or more SIDS deaths, premature African American males, and poor families with inadequate prenatal care.

TREATMENT   Apnea monitoring and family training for emergencies is crucial. Cardiopulmonary Resuscitation (CPR) and Basic Life Support (BLS) training should be required for family members in order to give the care necessary prior to emergency personnel arriving.

Note: CPR, we cannot imagine having a family and not becoming certified in CPR or BLS (see chapter 1). You can save a life, as they say GET ER DONE...

Swine Flu (North-American Influenza, H1N1, H1N3, Hog Flu)

DEFINITION   A mild to severe primary respiratory virus (first isolated in 1930 and first appeared in 1976 as a seasonal flu) affecting 1 in 6 Americans which is a strain of the influenza "A" virus subtype H1N1. It is thought to be a genetic combination of swine, avian and human influenza virus capable of spreading fro human to human via droplets (living on surfaces for 6-24 hours). H1N1 has formed at least two strains of the virus through mutation.

CAUSE   This deep respiratory virus infects the cell lining in your nose, throat and lungs. Cooked food products are safe and do not cause swine flu. Since humans have not experienced this strain of influenza we possess little natural immunity against the infection.

SYMPTOMS   Similar to other flu strains which symptoms develop 1 to 7 days after exposure (lasts for about 1 week) and remains contagious for about 8 days (24 hours before and after fever). This virus has displayed more serious symptoms in pregnant women and younger people. Symptoms include; fever ($\geq$100 degrees **for 3 or more days**), productive cough (colored sputum), sneezing, stuffy nose, sore throat, body aches, headache, chills fatigue, diarrhea, vomiting, difficulty breathing (dyspnea), pulmonary edema (extra fluids within lung tissue), chest pain/pressure, runny nose, malaise (tired), sudden dizziness, confusion and/or vomiting. These symptoms usually improve but often return with a fever or worse cough. In addition to symptoms listed above, children may exhibit rapid breathing (Tachypnea), trouble breathing (dyspnea), blue/gray skin color, severe/persistent vomiting, not drinking enough fluids, not waking/interacting and/or not wanting to be held. Patients with bacterial infections and/or chronic conditions (asthma, COPD, morbid obesity, immune deficiency etc.) are at increased risk of severe symptoms and/or death. Children with sickle cell disease are at greater

	risk of developing acute (quick onset) chest syndrome and may require intensive care.
TEST	Your healthcare providers will ask where you have traveled and may order a chest X-ray in order to look for pneumonia progressing in your lungs. Your healthcare provider may order influenza screening tests (i.e. Rapid Influenza Diagnostic Tests (RIDT), Nucleic Acid Amplifications Tests (rRT-PCR), viral culture (Petri dish), Direct and Indirect Immunofluorescence Assays (DFA & IFA), etc.) to differentiate swine flu from other influenza "A" subtypes.
TREATMENT	Prevention is the best bet, wash your hands, eat well, avoid infectious situations and follow your healthcare provider's recommendations for the Swine Flu vaccine. The vaccine is available as an injection/shot (nonliving form) and nasal spray (which is a weak live form of the virus). **Children 9 years old and younger require two (2) doses over a three week (21-28 days apart) period** (vaccine is not recommended for children less than 6 months old). H1N1 has mutated to H1N3 and will continue to mutate making it resistant to specific mediations for treatment. Often the flu is a self-limiting illness (you recover on your own) so drink plenty of liquids, consider pain relievers (analgesics) and rest. Never allow children less than 18 years old use aspirin (ASA) due to the risk of Reye's syndrome. Depending on the type of flu and other medical factors (i.e. age, health profile etc.) your healthcare provider may prescribe an antiviral medication or a combination of antiviral medications for 5 to 10 days. These medications include Neuraminidase inhibitors oseltamivir (Tamiflu or zanamivir (Relenza) and Adamantanes like amantadine or rimantadine. The Food and Drug Administration (FDA) has approved the emergency use of the IV (intravascular) Antiviral Peramivir (a neuraminidase inhibitor like Tamiflu or Relenza) for hospitalized children and adults. If H1N1 is accompanied with a bacterial infection or pneumonia (about half the cases) your healthcare provider will treat the bacterial infection with antibiotics. Prevention is your best bet; wash your hands frequently, try not to touch your nose, eyes or mouth, do not share food, drink or eating utensils, and discuss one of the vaccines available with your healthcare provider (vaccines take up to 3 weeks in order to develop immunity).

NOTE: Never take an antiviral as a preventative action unless your healthcare provider prescribes them and DO NOT buy antiviral over the internet. As a matter of fact never buy any medication over the internet without consulting with your healthcare provider.

Systolic Heart Failure (SHF) (see Heart Failure)

Transient Ischemic Attack (TIA) (see Stroke)

DEFINITION	This temporary blood vessel blockage leaves little to no permanent damage within the brain.
SYMPTOMS	Symptoms will be quick to appear and go away quickly and may include facial weakness, loss of coordination, loss of balance, visual/speech impairment and or headache.

Note: TIA symptoms are the same as a stroke but the symptoms must last for 24 hours to actually be called a stroke. A severe migraine or a migraine with aura (visual disturbances such as flashes of light, blind spots, light patterns etc.) can mimic the symptoms of a stroke. A TIA is a medical emergency and should be considered a warning. Call 911 or see you healthcare provider immediately.

Transient Tachypnea of the Neonate (Type II RDS or TTN, or Wet Lung) NEONATAL

DEFINITION   The starting and stopping of rapid breathing.
CAUSE   The delay in clearance of fetal lung fluids results in increased airway resistance and decreases lung compliance (stiffer lungs) resulting in hyperinflation (over inflation) and air trapping (airways close not allowing the air to escape the lungs).
SYMPTOMS   During the first few hours respiratory rate increases, pH (acid base balance) and CO2 (carbon dioxide) are normal.
TEST   Chest x-ray mimics pneumonia, perihilar streaking (white fuzzy lines) due to lymphatic engorgement (extra fluid in the lymph system), possible pleural effusions (fluid around the lungs and between the chest walls) in the Costophrenic angles (right and left tips of the diaphragm) and interlobular fissures (see chapter 7).
TREATMENT   Supplemental oxygen at the lowest level possible via an Oxyhood (looks like half a fish bowl placed over the head), nasal cannula or CPAP (Continuous Positive Airway Pressure). The patient's position should be changed frequently to help increase fluid clearance, which should happen within 24 to 48 hours. Your healthcare provider may order antibiotics and in severe cases, mechanical ventilation may be required however this is rare.

TTP (Thrombotic Thrombocytopenic Purpura or Thrombocytopenia)

DEFINITION   A rare disorder of the blood which results in a very low number of platelets in the blood (the cells that make the blood clot).

Note: This disorder can result in organ damage.

Tuberculosis (TB, Mycobacterium Tuberculosis)

DEFINITION   A chronic granulomatous (containing granulomas which are tumors or growths) resulting from infection caused by an acid-fast bacillus, Mycobacterium Tuberculosis.
CAUSE   Mycobacterium Tuberculosis (MTB) is a bacterial infection caused by the inhalation of contaminated airborne droplets characterized clinically by a lifelong balance between the host (you) and the infection. Pulmonary tuberculosis is an inflammatory reaction forming tubercles made up of giant cells and epithelia cells, which causes fibrosis (unusual fiber growths) and increases the probability of necrosis (dead tissue).
SYMPTOMS   Patient's are often asymptomatic (no signs) but the infection can be fatal. The first indication of TB is a positive PPD skin test or a positive AFB sputum test (requires three consecutive positive tests with three different first morning samples). The patient may have a cough with

169

mucopurulent sputum (mucus and pus), hemoptysis (bleeding from the lungs), dull chest pain, dyspnea (shortness of breath or difficulty breathing), tachycardia (rapid heart rate >100), fatigue, weight loss, night sweats, fever, and/or enlarged lymph nodes.

TESTS   Acid-Fast Bacillus (AFB) is laboratory tests using first morning sputum samples on three separate mornings to detect TB (all three must be positive) and is required by law. Chest x-ray may reveal infiltrates (white spots) or consolidation (looks like pneumonia) usually in the apical segment (see lung anatomy in chapter 7).

TREATMENT   Treatment consists of long-term drug therapy with isoniazid (INH) and/or rifampin (RMP), streptomycin (SM) and ethambutol (EMB), for months at a time. Patient education in effective pulmonary hygiene (moving secretions) is critical.

Tularemia (Rabbit Fever, Deer Fly Fever, Pahvant Valley Plague, Ohara's Fever)

DEFINITION   A rare infection consisting of several strains of bacteria carried by rabbits, beavers, muskrats and rodents.

CAUSE   The bacteria is normally spread by deer flies, ticks and most arthropods (i.e. things with exoskeletons or no bones inside their bodies). The Center for Disease Control considers the bacteria as a viable bio-weapon agent for bioterrorism.

SYMPTOMS   Symptoms are dependant on the location of the infection (there are six clinical syndromes). Incubation is 1 to 14 days which cause fever, lethargy (tired or lack of energy), weight loss, skin lesions (sores), swollen lymph nodes, swollen/redden eyes and face, and possible death.

TESTS   The laboratory will us blood samples to culture or grow samples for diagnosis.

TREATMENT   Your healthcare provider will prescribe an antibiotic (i.e. Streptomycin, Gentamicin, doxycycline etc.) for 10 to 21 days. A vaccine is available however it is restricted to use in high risk groups.

Tumors of the Heart (see Heart Tumors)

Tumors of the Lung (see Bronchogenic Carcinoma / Lung Cancer)

Valvular Heart Disease (Heart Valve Disease)

DEFINITION   A disease that affects one or more of the four main valves of the heart which can lead to heart failure.

CAUSE   There are several types of heart valve diseases; Congenital Valve Disease (malformed from birth) like Bicuspid Aortic Valve Disease (two instead of three leaflets, Valvular Stenosis (narrowed, stiffened or fused valve leaflets), Valvular Insufficiency (leaky valves or weak valves), and Mitral Valve Prolapse or MVP (the valve leaflets bend back into the left atrium) to name a few.

SYMPTOMS   Irregular pulse (arrhythmias or palpitations), chest pain, fatigue, swelling of the feet or ankles or abdomen (edema), rapid weight gain, shortness of breath (dyspnea), nausea, difficulty lying down to sleep and/or dizziness.

TEST   Your doctor will perform a physical exam and other tests. Often your
                doctor can hear a swishing sound (murmur) and hear any fluid build up
                (edema) in your lungs. Special tests such as echocardiography (see
                chapter 11), Cardiac Catheterization or angiogram (see chapter 11), MRI
                (Magnetic Resonance Imaging, see chapter 11) and Transesophageal
                Echocardiography (see chapter 11).
TREATMENT   There are three goals; reducing your symptoms, preventing further
                damage and repairing or replacing the valves which depends on the
                severity and type of valve disease.

Note: You must inform your dentist that you have any heart valve disease so that you are
protected from endocarditis when having dental procedures.

Ventilator-Associated Pneumonia (see Pneumonia)

DEFINITION   Ventilator-associated pneumonia (VAP) is one of the most common
                causes of hospital-acquired infections (Nosocomial Infections) for
                patients admitted to intensive care units.
CAUSE   Bacterial contamination in the lower airways often develops in intubated
                mechanically ventilated patients after 48 hours. Risk can be reduced by
                elevating the head of the bet to 30 degrees, proper oral care (often as
                ever 4 hours), brush teeth ever 12 hours, careful monitoring of feeding,
                minimizing ET Tube movement, and suctioning above the ET Tube
                balloon seal (to prevent aspiration, contamination to lungs from bacteria
                above the ET Tube balloon).

Note: Hand washing is the best way to reduce almost all cross contaminations.

Venous Thromboembolic Disease (see DVT and Pulmonary Embolism)

**Vocal Cord Dysfunction (VCD) also called (Paradoxical Vocal Fold Motion (PVFM) or
Vocal Cord Asthma)**

DEFINITION   A rare disorder which is characterized by full or partial vocal fold
                (voice box) closure or the vocal cords being closed over the airway that
                usually occurs during inhalation or exhalation for short periods of time.
CAUSE   The cause is not fully understood, the episodes can be triggered suddenly
                or have a gradual onset. It is predominantly observed in females 20-40
                years of age, however, can occur in people from 6 to 83 years of age.
                VCD is normally only experienced during waking hours and has been
                linked with extreme participation in competitive sports and in families
                with high achievement drives. It is believed that occupational irritant
                exposure to agents such as glutaraldehyde, strong odors (perfumes) and
                chlorine as well as exposure to inhaled environmental irritants, allergens
                and post nasal drip can cause an attack. Other medical conditions such
                as; Gastroesophageal reflux disease (GERD), Extra-esophageal reflux
                (EERD), psychological stressors (i.e. depression, stress, anxiety, mental
                disorders, neuroses, obsessive compulsive disorder) and exercise have
                been liked to the disorder.

SYPTOMS   Symptoms include; hoarse voice or inability to speak and/or other symptoms which mimic asthma, anaphylaxis, collapses lung, pulmonary embolism or fat embolism which occur while you are awake. Symptoms also will include; stridor (sounds like a barking seal), shortness of breath (dyspnea), wheezing, coughing, tightness in the throat/neck/chest and discoloration of the skin (blue or grey color leading to loss of consciousness in severe cases) due to oxygen deprivation. Additionally, signs and symptoms resemble conditions observed in disorders such as vocal cord paralysis, epiglottitis and laryngospasm. The usual asthma inhalers and breathing treatments will not relieve the symptoms.

TESTS   Pulmonary Function Testing (Exhaled Nitric Oxide Levels, Flow-volume loops and Spirometry) obtained during periods in which you are experiencing difficulty breathing (a symptomatic period) will indicate a limitation of inspiratory (inhaling) flow detecting the obstruction and eliminating the airways in the lungs as a suspect. The best test for a definitive diagnosis is directly observing the movement of the vocal cords using a nasal endoscope (a small camera) at the end of a long thin tube. A neurological evaluation is warranted to rule out any neurologic cause (i.e. cerebral palsy, brain stem compression, etc.) for the breathing difficulty. The observation that little or no response to beta-agonists or inhaled corticosteroids, a detailed medical history, nasal endoscope (a small camera), chest X-ray (to illuminate lung problems) and Pulmonary Function Testing are used to diagnose the condition. It is also crucial that your healthcare provider be suspicious of this rare condition because test results can only be definitive during an attack.

TREATMENT   You may find that by sitting down and taking slow, deep breaths or starting breathing exercises the condition will gradually resolve all of the symptoms in a few minutes. Your healthcare provider will stop any unnecessary asthma related medications (due to their severe long-term consequences such as growth retardation) and begin speech therapy. He/she may prescribe Botox injections, heliox (oxygen mixed with helium), Intermittent Positive Pressure Ventilation (IPPV), Continuous Positive Airway Pressure (CPAP), breathing exercises and/or tracheotomy (in severe cases). Studies indicate that VCD often resolves itself within in 5 to 7 years.

Whooping Cough (see Pertussis)

Notes:

Chapter 10
Medications

Medications, so many, so very confusing! Most people have more than one doctor treating various problems. There is nothing wrong with this as long you use only one pharmacy. You should always make sure each physician and pharmacy all have a list of every medication you are prescribed, including any over the counter (OTC) herbal/natural medications. By having only one pharmacy supplying all of your prescriptions you will have a second defense for alerting you to medications that may not go well together. If you use mail order pharmacies in addition to your local one, they too should have listings of all your medications.

One should keep an all-inclusive list of medications on their person at all times as well as with one's significant other. Trusted people who will have an input on your medical care should also have a list of your medications. A copy of these lists as well as any legal documentation (which will be discussed in chapter 4) should be kept on the refrigerator for ambulance personnel.

Now let us talk about "Emergency Medications" or Rescue Medications. They are difficult to identify even for the person who uses them. Could anyone in your family identify your emergency inhaler?

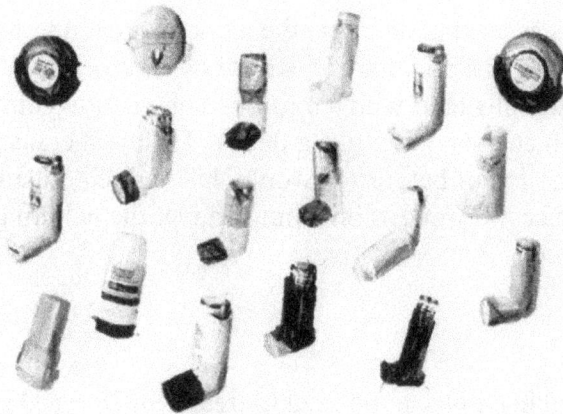

Inhalers

In a respiratory emergency, which one do you use or which one do you hand to mom or dad? The best way to identify the correct one is to mark it or place it in a separate baggy marked "Emergency Inhaler". A fast acting "rescue medication" is needed in an emergency. Using an improper medication will not help the situation, may make the situation worse or may even cause an overdose. These are important concepts that may prevent a trip to the emergency room or worse.

Now, let us look at medications, what they do for us and how they are used.

NOTE: New medications are continuously being developed and current recommendations or warnings for use and storage also change. The most current information on all medications starts with your doctor and pharmacist. Additional information about your medications can be found on the US Food & Drug Administration website (http://www.fda.gov) or on the manufacturing company's website.

Medical Gases

Oxygen therapy is designed to ensure your blood maintains the necessary level of oxygen. Supplemental oxygen is used to treat hypoxemia (low blood oxygen level), to decrease the work of breathing and to decrease myocardial (heart muscle) work. The testing for this is discussed in Chapter 11 under Diagnostic Testing in Arterial Blood Gas (ABG).

The finger probe that glows red called a saturation (sat) monitor will give an accurate reading as long as your hands are warm, no tremors and your circulation is adequate. It is interesting to note that the probe for the pulseox can be place almost anywhere ear, nose, forehead, toe and many other places on the body.

The human body likes everything in a status quo or in medical terms "homeostasis". When anything is not within the acceptable limits, the body will take steps to get things back to normal. If your oxygen level is too low, your heart and breathing rates will increase in order to provide the oxygen level necessary to every cell in the body. Most people "desaturate" (decreases in blood oxygen level) at night while sleeping. The reason for this is that while asleep one tends to breath shallower and less often. As a result, it is quite common for individuals to require oxygen at night, during sleep (see chapter 11, Sleep Study).

Helium Therapy

Helium is a low-density gas used to decrease the work of breathing when the patient has obstructions the airways (i.e. asthma, COPD, stridor or croup). Usually the mixture of oxygen and helium are administered with a mask, which distorts your voice temperately (like inhaling a helium balloon and sounding like a duck). Different concentrations are used and would be ordered by your doctor but, rarely would this be used outside the hospital setting. The whole idea is to reduce the work of breathing and let the helium get the oxygen down in the lungs where it is needed.

Nitric Oxide (NO)

Nitric Oxide gas (or laughing gas) can be used to treat COPD, ARDS, hypoxic respiratory failure, pulmonary hypertension and in lung transplants. Nitric oxide is a pulmonary vasodilator (similar to a nitroglycerin tablet taken for chest pain) that increases oxygenation and is normally not done outside the hospital setting.

General Medication information

Drugs are any chemicals that alter an organism's functions or processes. These chemicals come from plants, animals, and some are synthetic. Drugs are absorbed quickly through large surface areas such as the lungs however very few medications can be administered via the lungs. They are named by their brand name, generic name or chemical name. This can be very confusing, so do not be afraid to ask your healthcare professional for clarification.

Pharmacokinetics is the study of the absorption, metabolism (how the body breaks down the medications) within the human body, and the excretion of drugs from the human body.

Metabolism is the chemical process carried on within the body to sustain life. Your liver is responsible for the metabolism of medications. After metabolizing the medications, they are eliminated from the body via the kidneys or through dialysis.

Routes of Medication Administration

Most people use only oral, under the tongue, inhalation and injection. Other routes listed for your information.

Inhalation by using a nebulizer, MDI (metered dose inhaler), and DPI (dry powder inhaler) are the most common and are very fast acting.
Oral by swallowing a drink, capsule, or tablet pill take time to be absorbed by the digestive track and are slow acting (approximately 45 minutes).
IC (Intracardiac) is the fastest, but means the medications are injected directly into the heart.
IV (Intravenously) or into the vein has a very rapid effect and normally emergency medications requiring administration skills.
SL (Sublingual) or under the tongue is also very rapid (i.e. nitroglycerine tables etc.) slower than IV or IC but faster than the rest.
Intratracheal through an endotracheal tube (into the lungs directly via an ET Tube, see chapter 6) requires more medication volume (2-3 times the normal amount) in emergencies.
SQ (Subcutaneous) or under the skin, medium speed, slower than IV and faster than oral.
IM (Intramuscular) or into the muscle, medium speed, slower than IV and faster than oral.

Respiratory Medications

Here are well over a thousand medications associated with Respiratory Therapy. Below we will discuss the most common mediations used by most Respiratory Therapy patients.

Medication	Neb	MDI	DPI	Other	Type	Information
Accolate				Oral	7	zafirlukast
Advair		X*	X		1+ 3	fluticasone & salmeterol
Aerobid	X	X			3	flunisolide
Albuterol	X	X		Tablet & Syrup	1	Proventil HFA, ProAir, Ventolin
Alupent	X	X*		Tablet & spray	1	Metaproterenol
Alvesco		X		Tablet & spray	3	ciclesonide
Aminophylline				IV & oral	5	
Asmanex			X		3	mometasone furoate
Atrovent	X	X		Nasal	1	ipatropium
Azmacort		X			3	triamcinolone acetonide
Beclomethasone		X		Nasal	3	Qvar
Brovana	X				1	arformoterol tartrate

Combivent		X			1	Alabuterol & ipratropium bromide
Cromolyn Sodium	X	X		Spinhaler & Nasal		Intal, anti-inflammatory for bronchail asthma
Dexamethasone		X		Oral, Drops, Topical & Injection	3	
Dornase	X				2	pulmozyme
Dulera		X			1 + 3	mometasone & formoterol
Duoneb	X				1	Alabuterol & ipratropium bromide
Epi Pen				Injection		epinephrine for allergic reactions
Flovent		X	X		3	fluticasone propionate
Foradil		X			1	formoterol fumarate
Gentamicin	X			Injection, Drips & Topical	4	
Maxair		X			1	pirbuterol acetate
Methylprednisolone				Tablet, Liquid & IV	3	
Acetylcysteine	X	X		Spray & Tablet	2	Mucomyst or Mucosil (old names)
Nedocromil Sodium		X *				Stabilizes mast cells in hypersensitivity reactions
Pentamidine	X	X			4	
Pulmicort	X	X	X		3	budesonide
Quvar		X			3	beclomethasone
Racemic Epi	X					
Ribavirin	X			Oral & Injection	2	Large Volume Nebulizer is used
Saline 0.9% & 3%	X				2	salt water
Serevent		X *	X		1+ 3	salmeterol xinafoate
Singulair				Oral	7	montelukast
Sodium Bicarb	X					
Spiriva			X		1	tiotropium bromide
Symbicort		X			1 + 3	budesonide & formoterol fumarate
Terbutaline				Injection or Tablet	1	

Theophylline				Oral, Injection	5	
Tobramycin	X				4	
Xolair				Injection	6	omalizumab
Xopenex	X	X			1	levalbuterol
Zyflo				Oral	8	zileuton

* Medications are being discontinued or phased out.

Note: Many medications have similar packaging. Delivery systems and propellants (i.e. MDI, DPI, CFC, HFA etc.) can be confusing (see chapter 6 Hand Held Inhalers). The FDA warns you to read the container carefully in order to identify the proper medication.

1. Bronchodilators

Bronchodilators affect the body's nervous systems by stimulating one or more receptors. Beta 1 (remembered by thinking 1 heart) increases the heart rate and increases the strength of cardiac contractions. Beta 2 (remembered by thinking 2 lungs) acts as a direct bronchodilator of the lungs, stimulates mucociliary (mucus) activity and has a weak Vasodilatation (relaxes the venous walls) effects. In the world of respiratory, the Beta 2 works better for breathing and has little or no effect on the heart or vessels. Some bronchodilators are short acting and some are long acting. This is why your healthcare provider will prescribe how often they should be taken.

Caution: Allergies to soy products or peanuts may be an indication you may have an adverse reaction to Atrovent or ipratropium bromide and Spiriva or tiotropium bromide. Additionally care should be used if you have narrow-angle glaucoma. Talk to your healthcare provider or pharmacist about any risks with these medications.

Note: Warm coffee and some tea's act as a bronchodilator. This may be why smokers love that morning cup.

2. Mucolytics and Wetting Agents

Mucolytics help break up thick mucus and making it easier to expel through breakdown the disulfide bonds that hold the mucus together. Wetting agents are used to thin out the mucus using liquids.

Use caution when using mucolytics in the elderly with asthma. Mucolytics should not be nebulized in the same nebulizer (without cleaning). Dornase should be nebulized alone. Additionally, mucolytics can cause mild dehydration, so talk to you healthcare provider concerning additional fluid requirements.

3. Corticosteroids (Steroids)

Corticosteroids (Glucocorticoids) are used to treat inflammatory pulmonary diseases, including asthma and COPD. Corticosteroids reduce inflammation by stabilizing leukocyte in the lysosomal membranes, suppress immune responses of pulmonary mucosa to allergens and enhance the effectiveness of bronchodilators. Corticosteroids can take 1-4 weeks to reach full effectiveness.

Note: Always rinse mouth after ingesting steroids or using steroid inhaler to prevent thrush (they are the white mouth sores babies get).

Adverse Reactions include headache, dizziness, seizures, insomnia, nausea, weight gain, water retention (edema), Cushingoid state (Cushing's disease), muscle wasting, arrhythmias (abnormal heartbeats), bronchospasm (closing of airways), cough, hoarseness, sore mouth (Thrush or Candidiasis). Discuss this with your healthcare provider or pharmacist.

4. Antibiotic Medications

Antibiotic medications are given via nebulizers to treat bacterial infections within the lungs. Always take these medications as prescribed. Never stop taking them because you feel better, this can cause the bacteria to develop resistance to the medications. Antibiotics should not be combined with mucolytics in order to nebulize them together.

5. Xanthine drugs relax the bronchial airways and pulmonary vasodilatation. These are not common medications.

6.   Immunoglobulin E (IgE) is responsible for asthma attacks. These medications reduce the severity and frequency of the attacks (immune system).

7.   Long term leukotriene receptor antagonist (block the stuff that causes asthma attacks).

8.   Leukotriene inhibitor (inhibits the stuff that causes asthma attacks).

Neuromuscular Blocking Agents

These medications are never used outside the clinical setting with the exception of an ambulance. Neuromuscular blocking medications (i.e. Succinylcholine, Pancuronium, Tubocurarine, etc.) induce skeletal muscle relaxation or paralysis to facilitate intubation and to assist with mechanical ventilation or other medical procedures.

Surfactant

Surfactant is lubrication within the lungs. These medications are used to prevent and treat IRDS (Infant Respiratory Distress Syndrome see chapter 9) administered directly into the lungs. Surfactants for treatment of adult ARDS, pneumonia and other diseases are under investigation.

Cardiac Medications (Heart Medications)

Antiplatelet Agents

Used to prevent blood clots in the blood by preventing blood platelets (adhere together to stop bleeding) from sticking together (aspirin (ASA), Clopidogrel, Ticlopidine and others).

Blood Thinners (Anticoagulant & Antiplatelet)

Blood thinners are used to thin your blood (slow coagulation or prevent clotting) in order to prevent harmful clots from forming in your blood. Medications used are Coumadin or Warfarin, Heparin, Dalteparin, Enoxaparin and others slow coagulation. Medications like Aspirin prevent clots. It is important to note that green vegetables alter the dosing. This is a good topic for discussion with your healthcare provider and pharmacist.

ACE Inhibitors (Angiotensin-Converting Enzyme Inhibitors)

Medications used to treat heart failure and high blood pressure (hypertension) by expanding the blood vessels and by lowering Angiotensin II (a powerful vasopressor produced due to the kidneys). This allows the blood to flow through the body with less resistance and reduces the work of the heart (Lotensin, Captopril, Enalapril, Lisinopril, Ramipril, Quinapril etc.).

Angiotensin II Receptor Blockers (ARB's)

Used to treat heart failure and high blood pressure (hypertension), these blockers or inhibitors prevent or block the vasopressor effects on the blood vessels in order to reduce blood pressure. Medications include Losartan, Candesartan, Irbesartan, Valsartan and others.

Beta Blockers (Beta Adrenergic Blocking Agents)

Used to treat tachycardia (rapid heart rate >100), lowers blood pressure and treats chest pain (angina). These medications decrease heart rate and cardiac output, which in turn reduces blood pressure. Medications include Atenolol, Hydrochlorothiazide, Carteolol, Metoprolol, Propranolol and many others.

Calcium Channel Blockers (Calcium Antagonists)

Used to treat high blood pressure (hypertension), chest pain (angina) due to reduced blood supply to the heart muscles by blocking the movement of calcium. Additionally they are used to treat some arrhythmias (abnormal heart beats). Medications include Diltiazem, Nifedipine, Verapamil, Isoptin and many others.

Diuretics (Water Pills)

Used to treat fluid over load or excess fluids (hypervolemia), high blood pressure (hypertension), pulmonary edema (liquid in the lung walls), and general edema (such as swollen ankles or feet). These medications remove extra fluids through the kidneys and reduce the workload of the heart. Medications include Furosemide, Hydro-chlorothiazide (HTCZ), Amiloride, Bumetanide and others.

Note: Your doctor may add potassium to your medication list while you are using diuretics.

Vasodilators (lowers or reduces blood pressure)

Used to treat chest pain (angina) by relaxing blood vessels and increasing blood flow to the heart muscles while reducing the workload of the heart due to back pressure in the veins. Additionally these medications may be prescribed to patients who cannot tolerate ACE Inhibitors. Medications include Nitroglycerin, Nitroprusside, Nesiritide, Minoxidil, and others.

Vasopressors (increases blood pressure)

Used to treat hypotension (low blood pressure) through the alpha response that results in vasoconstriction of the veins and arteriolar smooth muscles this will increase blood pressure (think of it as squeezing a balloon). Medications include Dobutrex, Dopamine or Intropin and Norepinephrine.

Cardiac Glycosides

Medications used to treat CHF (Congestive Heart Failure) and arrhythmias (abnormal heart beats) by increasing the heart's contractions and relieve heart failure symptoms. Medications include Digoxin, Inamrinone, Milrinone and others.

Analgesics (Pain Relievers)

Used to relieve pain, produce sedation, anesthesia, muscle relaxation, anxiety relief, and/or to aid in sleep. Most sedatives are Central Nervous System (CNS) depressants and can cause respiratory depression (shallow, slow breathing or even cessation of breathing).

Narcotic Analgesics are used for pain relief and sedation. Fentanyl or Sublimaze, Demerol, or Meperidine, Darvon or Propoxyphene, and Morphine or MS Contin are just a few.

Sedatives/Tranquilizers are used for anti-anxiety and sedation. Benzodiazepines such as; Diazepam or Valium, Lorazepam or Ativan, Chlordiazepoxide or Librium, and Propofol or Diprivan to name a few.

Analgesics and sedatives can become habit forming. Discuss alternatives with your healthcare provider or pharmacist.

This is a very small sample of medications; it would require up to 2000 pages just for medications alone however, we have covered the important respiratory and cardiac related ones. More information about these medications and many others can be found at http://www.fda,gov, http:///www.webmd.com, or other sources (see chapter 1).

Nebulizer medication compatibility (the ability to mix medications together) a very important issue, so we will review it here. Always follow your healthcare provider and pharmacy instructions to the letter. Dornase should always be nebulized separately from your other medications in a second hand held nebulizer (or at least in a freshly cleaned nebulizer). Antibiotics should not be combined with mucolytics in order to nebulize them together. Tobramycin should be given in a separate nebulizer when using budesonide or cromolyn. Never mix Brovana with any other nebulizer medications.

Just a few reminders; (1) never take anyone else's medication, (2) always read the package inserts for each medication you are prescribed, (3) tell you doctor and pharmacy about all medications you use, include herbs, and (4) maybe the most important, don't modify your medication schedule without consulting your healthcare provider or making him/her aware of your changes.

Be well and be safe...

Notes:

Chapter 11
Special Testing and Surgical Procedures

Surgical and non-surgical procedures have come a long way over the past 100 years. Survivability and the resulting quality of life have improved greatly. Across the United States, thousands of people undergo heart and lung surgeries every day. Most of these patients will live longer and have a much better quality of life; however, although surgical outcomes have tremendously improved, it is seldom a bad idea to get a second opinion before undergoing any procedure. We have selected the most common types of surgeries/procedures and have placed them in alphabetic order, as you will see below:

Note: You may want to review Cardiac Catheterization because it applies to many of the procedures. You may also find that may of the procedures are very similar sounding but, they all have their specific requirements, equipment and are performed for very different reasons.

Advanced Life Support Procedures (the basics)

Advanced Life Support (or Pediatric Advanced Life Support in the case of young patients) is part of the training Paramedics, Respiratory Therapists, Nurses and Doctors receive to save lives. It includes everything from births to traumatic injuries and can include surgical procedures. Everyone has heard of the **"Golden Hour"**. That's the 60 minute period after injury your medical team has to save your life, it expands on the term "Time Is Muscle" for your heart. It is up to you and your loved ones to call for help, the clock has already started.

The advanced Cardiac Life Support (ACLS) algorithms (step by step procedures) include assessments airway (windpipe), breathing, circulation (pulses) and defibrillator (delivering shocks) or diagnosis, similar to chapter 9), Cardiopulmonary Resuscitation (CPR in chapter 1), Intubation (in chapter 6), Electrical Shock (the defibrillator which can monitor your rhythm or EKG) and Cardiac Medications.

In chapter 7 and 8 we learned that your heat is an electromechanical pump. CPR (Cardiopulmonary Resuscitation) can keep the blood pumping through your body however; an electrical shock from a defibrillator to reset and restart your heart is required. Defibrillators have two basic models, the professional version and the Automated External Deliberator you have seen in the mall or ballpark. Medications (about 25 different ones) help to establish and keep your heart in a normal sinus rhythm (as discussed in chapter 8 & 9).

Angioplasty (Balloon Dilation or Balloon Angioplasty)

A small balloon threaded into a coronary artery (supply the blood to the heart muscle) through a catheter. The balloon is then inflated to compress the plaque in order to restore the blood flow feeding the heart (also see Stent Procedure). First performed in 1977.

Note: The balloon dilation discussed above can also be done inside the nasal passages in order to help permanently open them and ease breathing through the nose.

Angioplasty Laser

Similar to angioplasty using a catheter equipped with a laser-emitting tip. The pulsating laser beam vaporizes the plaque build up in the coronary arteries (blood supply to the heart muscle).

Arterial Blood Gas (ABG)

The ABG is drawn from the wrist (radial artery), the fold of the elbow (brachial artery), near the pubic bone (femoral artery), or from an arterial line (see arterial line) that has been placed due to the requirement for frequent testing (also used for balloon pumps or mechanical ventilation patients).

You cannot see an artery like you can a vein, but you can usually feel the pulse with ease. The correct procedure to obtain an Arterial Blood Gas (ABG) should be similar to as follows: Gloves are required for your protection as well as that of the person performing the procedure. When drawing from the radial artery (wrist), collateral flow (through the ulnar artery on the inside of the wrist) is checked using an Allen's Test. This ensures the ulnar artery is working well in supplying blood to the hand. The site is cleaned and the proper location is assessed. The needle is inserted at approximately 45 degrees and the arterial blood should start to flow immediately. Sometimes, the needle must be repositioned, but the goal is first stick. Pressure should be held for at least 5 minutes and should be held for 10 minutes if the patient is taking blood thinners (anticoagulants like aspirin, heparin, Coumadin, etc.). ABG's do not usually hurt any more than any other needle. This depends on two things, the proficiency of the healthcare professional performing the procedure and not hitting any nerves. Nerves are too small to be seen, and there are a lot of them on the inside of the wrist. Therefore, there is some luck involved.

Note: Band-Aids are useless after the ABG or blood is drawn. Pressure should be held for 5 to 10 minutes to stop the bleeding. Holding pressure will also reduce or eliminate any bruising. Many medical personnel are either too busy or do not understand that pressure is the key to bleeding and bruising.

Arterial Line

Arterial lines are similar to an IV placed into veins (returning blood to be oxygenated) except they are placed into an artery (oxygenated blood). They normally are placed into the radial artery (bottom and outside of the wrist), the femoral artery in the groin or as a central line (located in the chest) placed near the heart and lungs using a long catheter. These lines are used when Arterial Blood Gas (ABG) sampling is required routinely. They provide the arterial blood sample without having to stick the patient repeatedly to obtain samples for testing.

Artificial Heart Valve (Heart Valve Replacement Surgery)

This procedure is used to replace defective, diseased or abnormal heart valves with a healthy valve.

Atherectomy

Similar to angioplasty using a catheter equipped with a rotating shaver (or blade) which cuts away plaque build up in the coronary arteries (blood supply to the heart muscle).

Body Plethysmography

This is an enclosed chamber or box accurately measures asthmatic airway resistance and air trapping within the lungs. A test will gather information or to baseline prior to administering a bronchodilator (like albuterol). A significant change is 35% or greater increase for the patient baseline or a decrease in the lung residual volume and decrease in total airway resistance (see PFT or Pulmonary Function Test at the end of this chapter).

Bronchoscopy

This procedure allows the physical examination of the bronchi (the airways of the lungs) through a bronchoscope (a thin flexible tube). Bronchoscopes come in different styles and some can display what the doctor is viewing on a TV screen while others are more like looking through a microscope. In any case, the bronchoscope is long thin flexible tube (that can fit easily through a drinking straw) with optics and a small light source. The scope also allows the doctor to suction out debris and even to take a small sample of tissue if required. Prior to using the bronchoscope, your doctor will numb your throat and sedate you for your comfort.

Capnometer (Capnograph or Capnography)

Measures the waste gas CO2 (carbon dioxide using a finger probe and others use exhaled breaths.

Cardiac Bypass Surgery (Open Heart Surgery or CABG)

Coronary (crown) artery bypass surgery, coronary artery bypass graft surgery (CABG), heart bypass or bypass surgery are all the same thing. The procedure is performed to relieve angina (chest pain) and reduce the risk of death from coronary artery disease (CAD) caused by the narrowing or blockage in flow feeding the heart. Arteries or veins taken from your body (normally the legs) are grafted (placed around) around the blockages supplying the heart muscle (myocardium) with food and oxygenated blood. The surgery can be performed on a beating heart (using CABG) or a non-beating heart (requires equipment for cardiopulmonary bypass or heart lung bypass). Your surgeon will discuss with you the type of surgery that best fits your coronary problems. Surgery will normally take 3 to 5 hours depending on the number of bypasses necessary and the complexity of the surgery (first performed in 1967).

In a nutshell, using the veins and/or arteries removed from your leg (the normal location) will be used as a bypass from the aorta to just past the blockage in the artery or vein. All the procedure does, is provide a new blood flow path to the part of the heart that is starving for oxygenated blood.

Cardiac Catheterization (cardiac cath or coronary angiogram)

Cardiac cath is an invasive (entering something into your body) imaging procedure that shows how well your heart and arteries are working and tests for disease in about 30-45 minutes. The catheter (a narrow tube) is inserted through a hollow needle into a blood vessel normally in your arm or leg and is then guided by x-ray to your heart. A dye (contrast) is injected through the catheter to allow your healthcare provider to see your heart valves, coronary arteries and the four chambers of the heart. This test diagnoses CAD (Coronary Artery Disease), heart valve disease, aorta disease and others. Additionally, the procedure is used to open blocked arteries using a balloon (to compress the blockage) or to place stents (looks like a spring) to keep arteries open. This procedure will require that you be monitored for bleeding after the procedure for from anywhere from three hours to overnight (depending on your health and needs).

Cardiac Doppler Study (Echocardiography or Echo)

This test identifies changes sound waves during the Echo or Doppler procedure to view the blood flow to the heart, inside the heart and blood flow from the heart. It helps determine if the heart is pumping well and/or if any obstructions could be causing problems with the blood flow within the heart including valves and structures of the heart walls.

Cardiac PET/CT Scan

The Cardiac Positron Emission Tomography (PET) and Computed Tomography (CT) (90 minute procedure) combine two medical imaging techniques to show both the chemical function and structures of the heart, including its surrounding vessels (muscle, veins and arteries). Diagnosing Coronary Heart Disease years before you experience a heart attack may allow your healthcare provider to reverse the process early without surgical procedures.

A small amount of radioactive tracer (presently Rubidium 82) which emits subatomic particles called positrons is injected into your arm. The blood stream carries the material until it is deposited in the heart muscle where a special camera detects the positron emission and constructs a detailed picture of your heart (muscle, veins and arteries).

Cardiac Stress Test

These tests increase heart rate, blood pressure and check for changes in your EKG. There are two types of cardiac stress tests, the treadmill stress test and pharmacological stress tests. They are done as follows:

> During the regular treadmill stress test no drugs are given. In the stress Thallium test a radioactive isotope (fast acting) is administered, when a predetermined heart rate is reached and a nuclear scan is performed upon immediately upon completion. In the stress Myoview a radioactive isotope (slow acting) is administered when a predetermined heart rate is reached the nuclear scan is done within 24 hours. The stress echo test is a regular treadmill test where an echo test (sonar view) of your heart is completed when a predetermined heart rate is reached (must be done within 2 minutes).

Adenosine Thallium or Myoview pharmacological stress test uses adenosine to speed up your heart. After 2 minutes, Thallium or Myoview (fast acting) is given over 4 minutes. This dilates the coronary arteries (simulates vigorous exercise). The Persantine Thallium or Myoview test dilates the coronary arteries (simulating vigorous exercise) and at 6 minutes, adenosine or Myoview is given. The Dobutamine test is only performed in the cardiac cath laboratory. Medications are given via an IV drip. The drip is increased every 3 minutes (q3) until a predetermined heart rate is reached. If the Dobutamine does not cause your heart to reach the predetermined heart rate, Atropine is given via the IV.

Cardiomyoplasty

There is not much information about this procedure because it is still experimental and has not yet been approved for the public. Muscle tissue taken from the patient's back or abdomen and is placed around the heart. A special type of pacemaker is attached and results in an improved pumping ability.

Cold Air Challenge

Cold air (just below 0 degrees Fahrenheit) mixed with Carbon Dioxide (CO_2) is deeply inhaled for three minutes after the expected restrictions are reached. The cold air generator is removed and spirometry tests are done at 1 minute, 5 minutes, 10 minutes and 20 minutes. Once the testing is complete, the patient is given a bronchodilator in order to reverse the cold induced asthma attack.

Coronary Artery Calcification (CAC)

This type of scanning is used to rule out the presence of plaque buildup in the coronary arteries (the arteries that supply your heart with blood). After your doctor has performed a Cardiac Stress Test (running/walking on a treadmill) your doctor can determine if any arteries may be clogged with plaque. The next step is usually a cardiac catheter to determine how badly they are clogged. The CAC scan if normal can negate (eliminate) the necessity for cardiac catheterization, which is an invasive test procedure (only a small hole).

Cricothyrotomy (also called thyrocricotomy, cricothyroidotomy, laryngotomy)

An emergency procedure to establish an airway during any emergency that blocks the nose, mouth or trachea (trauma, swelling, foreign objects etc.). See Tracheotomy for the procedure.

CT Scan (Computed Tomography) / **CAT Scan** (Computed Axial Tomography)

The CT Scan is a newer and faster version of the CAT scan. Both are advanced x-ray procedures using tomography (three-dimensional images) like a globe of the earth were you can feel the mountains. The map of the lungs, coronary arteries (in order to detect plaque and blockages) internal organs and structures of the body makes it easier to locate abnormalities. Additionally, the CT is used in to accurately guide the placement of instruments, or treatments during procedures and is excellent for detecting acute (rapid onset) & chronic (long-term) changes. A contrast material (dye) is used to help identify features or problems.

Dipyridamole Stress Test

A medication called dipyridamole (persantine) is administered to patients who can not run on a treadmill or ride an exercise bike which generates the same effects on the heart and blood vessels.

Ejection Fraction (Ef)

Ejection fractions is the fraction of blood pumped out of your ventricle with each heart beat or the capacity at which your heart is pumping. The term ejection fraction applies to both the right (RVEF) and left (LVEF) ventricles (see chapter 7). It is understood, unless specifically stated that ejection fraction refers specifically to the left ventricle. There is a considerable amount of information for testing your hearts flow however, healthy individuals have ejection fractions from 50% to 70% depending how the percentage is being calculated. Ejection Fraction is commonly measured by echocardiography that measures the volumes of the heart's chambers and dividing it by the stroke volume. Other methods include a cardiac catheterization, magnetic resonance imaging (MRI), computerized tomography (CT) and nuclear medicine scan (uses thallium injection).

Electrocardiogram (EKG)

EKG's were discussed shortly in chapter 7 & 8, but now we will go a little more in detail. There are three basic types of EKG's most people will see at some point in life. The three lead EKG is used to quickly assess your heart, the 12 lead EKG to give a more detailed assessment of your heart and the portable cardiac event monitoring (worn for 24 hours and much longer in some cases). Usually your healthcare provider downloads the event monitor information at the office and some can be transmitted over your phone to your doctor's office. The 12 Lead EKG (really, only 10 leads total) is used to monitor the heart in much greater detail and used during stress testing (treadmill or chemical tests) on your heart. There are other types of EKG's used by cardiologists to monitor specific events within your heart (these will not be discussed here but can be found online). When looking for information on the internet it is important to ensure that the site is trustworthy and accurate. You should always double check the information by using more than one website.

Arrhythmias (unusual heartbeat) and nocturnal complex arrhythmias (unusual heartbeat at night) have been associated with Sleep Disordered Breathing (SDB) also known as Sleep Apnea (See Chapter 9).

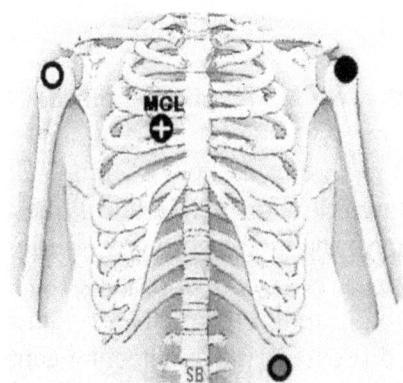

3 Lead EKG

The 3 lead EKG pads (electrodes) placed as shown above. This is a basic setup to monitor the heart during procedures or until a 12 Lead can be set up in its place.

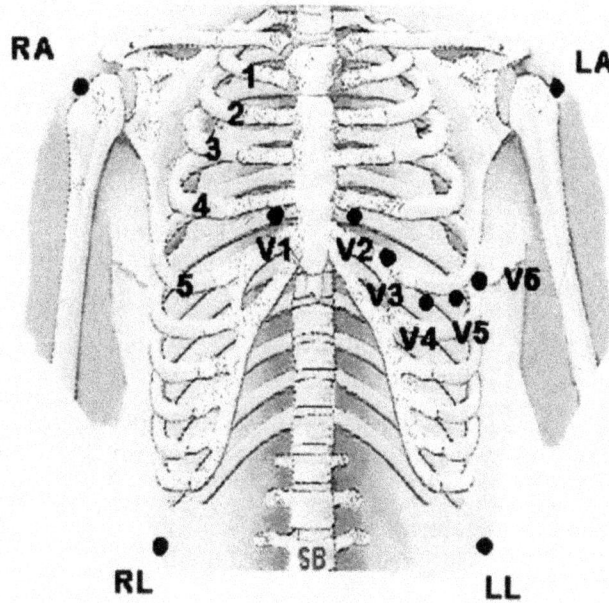

12 Lead EKG

The 12 Lead EKG leads are placed a shown above. This gives the doctor a detailed look at what the heart is doing electrically. One of the things your doctor is looking for is a history of ischemia (the reduction of blood flow to your heart causing damage to the heart muscle).

When your doctor orders a 12 EKG (electrocardiogram) this is what he or she looks at to determine how your heart is functioning electrically (see drawing below).

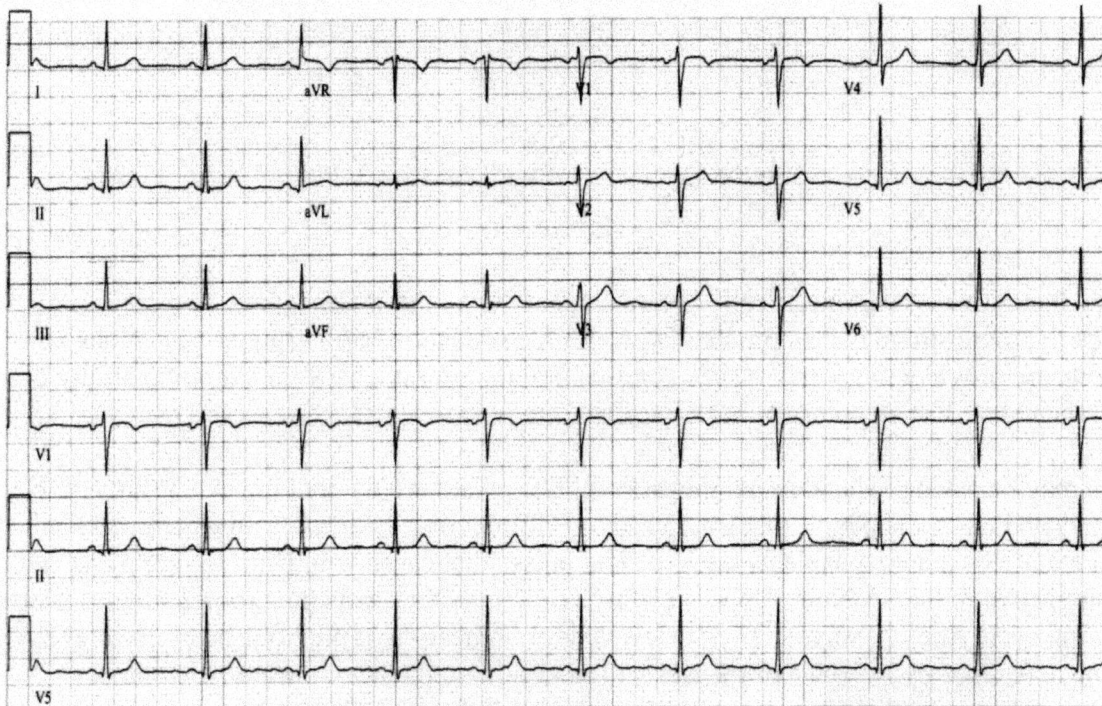

12 Lead EKG Trace

Below you see that each of the traces V1, V2, V3 etc. correspond to an area of your heart.

Coronary Arteries	Anatomical Structures Supplied	Corresponding EKG Lead (Trace)
Left Main	LAD & LCX	
Left Anterior (front) Descending LAD	Anterior wall of Left Ventricle, 2/3 septum, anterior Apex (top), Bundle branches	V1, V2, V3, V4
Left Circumflex (circling) LCX	Right atrium, lateral (side) wall of LV, posterior (back) wall of LV	I, aVL, V5, V6
Right Coronary Artery RCA	Right atrium, RV, inferior (lower) wall of LV, SA node in 55% of population AV node in 90% of population	II, III, aVF
Posterior (back) Descending Artery (from RCA or LCX)	Posterior wall of LV	Reciprocal changes in V1 & V2
Right & Left Bundle Branch Blocks RBBB		R obvious rabbit ears
LBBB		no R in V1 rabbit ears (see below).

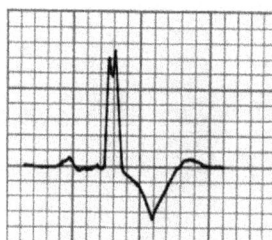

Rabbit Ears

Even though your heart electrically (heartbeats on the monitor) appears to be working properly, there still may be a problem with its mechanical operation (pumping). Your healthcare provider may order a special procedure in order to check the pumping ability of your heart Cardiac Doppler Study or Cardiac Cauterization).

You may be asking, what does all this mean? The pulses are not that important for you to understand but some people just want to know. Remember the heart is an electromechanical pump and this drawing explains the electrical pulses that cause your heart to pump and reset. A single cell in the SA Node (upper right side of your heart) starts the process of your hearts beating cycle (the "P" wave). The "Q" wave is an indication of previous heart muscle damage (caused by ischemia or the lack of blood flow to the heart muscle), but almost everyone has a small "Q" wave depression. Through the "QRS" complex (group of waves) is the heart beating and pumping blood throughout your body. During this complex is when you can feel your pulse (the systolic pressure or the higher number of your blood pressure) followed by the diastolic pressure (the lower number in your blood pressure, when your heart is resting). The "T" & "U" waves are reset pulses for your heart although normally the "U" wave is not normally seen on the EKG due to other electrical activity.

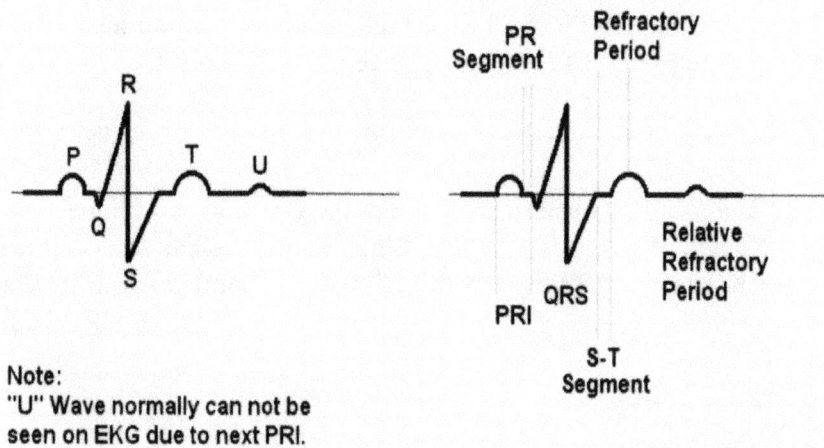

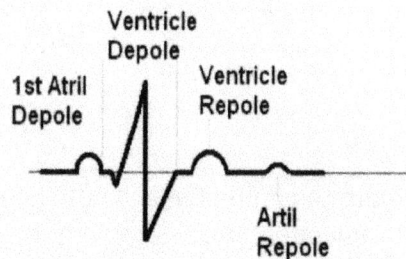

Electrical Pulses of the Heart

Electrocardiogram/ Electrophysiology EP Study

The EP study is a very detailed to evaluate the electrical pathways and activity of your heart. The test aides your healthcare provider in determine the cause of irregular heart beats (arrhythmias) and develop the correct care plan.

Electroencephalogram (EEG)

Measures and records brain wave activity (alpha, beta, delta and theta waves) in order to determine your stages of sleep. No needles are used.

Electromyogram (EMG)

Records muscle activity such as facial movements, twitches, teeth grinding, and leg movement (or Restless Leg Syndrome) during stages of sleep. No needles are used, movement sensors are attached with tape.

Electro-oculogram (EOG)

Records eye movement during stages of sleep (like REM). No needles are used, movement sensors are attached with tape.

Endobronchial Ultrasound (EBUS)

The Endobronchial ultrasound (a special bronchoscope) is a diagnostic tool allowing an ultrasound from inside the airways. It allows your doctor to see the airway walls, lymph nodes and structures surrounding the trachea as well as any abnormalities. The unit is also capable of removing tissue samples for analysis (see Bronchoscopy).

Event Monitors

The Event Monitor is a small medical device (about the size of a cell phone) that records your heart's electrical activities day and night. When you feel symptoms that may be cardiac event the recorded EKG information is transmitted over the phone to your heart specialist for review.

Exhaled Nitric Oxide Analysis

This test is an indicator of allergic reactions (or eosinophilic airway inflammation such as asthma) causing airway inflammation. When the airways are inflamed, it will take longer than normal to eliminate the nitric oxide from the lungs. This test also allows your healthcare provider to accurately wean (taper off and discontinue) you off prescribed steroids.

Gallium Scan

The test is a type of nuclear medicine exam that uses a radioactive material called gallium to look for swelling (inflammation), infection or cancer within the body. A special camera is used for the scan 6 to 24 hours after the injection of the radioactive material.

Heart Transplant

This procedure replaces a defective or diseased heart with a healthy human heart or maybe an artificial one. Patients must be appropriately selected and screened for this procedure and have over an 80 percent survival rate at three years. Due to the cutting of heart nerves, Atropine will have no effect on the new heart.

Heart Valve Surgery (Repair, Replacement or Balloon Valvotomy)

Heart valve surgery take between 3 to 6 hours depending on what procedures are required. Surgery Mitral Valve Regurgitation is the most common although Aortic, Tricuspid, Pulmonary and Bicuspid valves can also be surgically repaired or replaced.

A natural, homograft or allograft heart valve surgery is parts from human donors. In a modified natural or biological heart valve surgery parts come from animal donors (pig or cow) and parts are place in a synthetic ring for the proper shape and function. This type of surgery has a higher level of risk for infection. An artificial or mechanical heart valve surgery uses metal or pyrolytic carbon, which the body tolerates well. Silicone carbide is no longer used due to blood clots forming.

Heart valve repair is accomplished using mechanical parts tolerated well by the body, in the use of tissue donated by human or animal cadavers or surgically tightening the valve parts in order to make them function more efficiently.

Heart valve balloon Valvotomy is similar to Angioplasty (Balloon Dilation). A small balloon is threaded into the stenotic (narrow) valve. The balloon is inflated to stretch or open the valve in order to restore the proper blood. This procedure can be used on the Mitral Valve, Aortic Valve (Bicuspid) and when surgery is not an option for the patient can be performed for Pulmonic Stenosis (Pulmonary Valve).

Types of heart valve surgery:

Commissurotomy (widening the opening between valve leafs).

Decalcification (removing calcium deposit from the valve t restore flexibility).

Triangular Resection (also called chordal transfer, or reshape Leaflets) used to restore a flail (floppy) valve to make it stiffer.

Annulus Support (sewing a ring around the valve tissue to stiffen the valve).

Patched Leaflets (to repair tears and holes).

New heart valve surgical devises allow surgeons to repair or replace heart valves using a less invasive catheterization procedure (see cauterization).

A new mitral valve clip allows the surgeon to use a minimally invasive procedure (no open heat surgery required) to clamp the two flaps of the mitral valve together in order to prevent leaking (regurgitation). This procedure uses a catheter instead of open heart surgery which allows the patient to leave the hospital 1-5 days after the procedure (depending on the individual patient).

Holter Monitors

The Holter Monitor is a medical device (about the size of a cell phone) that records your heart's electrical activities. The Holter Monitor usually records EKG information for 24 hours, returned and downloaded by the heart specialist for review.

Hypoxic Altitude Simulation Test (HAST)

Patients use supplemental oxygen and plan to travel to altitudes of 5000 to 8000 feet or who plan to fly (cabin pressure is equivalent to 7000 to 8000 feet) should speak to the healthcare professional about having a HAST evaluation. This test should be performed with mild exercise (walking) and will determine if the individual can tolerate this type of travel without supplemental oxygen.

Implantable Cardioverter Defibrillator

These devices perform the same function as the defibrillators that you see on television or in the movies. They are used to shock a patient's heart and put it back into a normal rhythm. When required during ventricular tachycardia or ventricular fibrillation (fast and non-perfusing heartbeats, meaning that little or no blood are being pumped from the heart) the implant delivers an electrical shock, when the appropriate rhythm is detected the device returns to standby and in some cases it can pace the heart if required (see Pacemaker). This equipment can prevent sudden cardiac death.

Laryngectomy

The removal of the larynx (voice box) which usually disconnects the mouth and nose from the trachea results in an obstructed airway. An opening is left in the neck (stoma) which is used for breathing. Humidification may be required in order to maintain healthy lung tissue (the nose would normally provide the needed humidification).

Left Ventricular Assistance Device (LVAD)

The left ventricle pumps blood from the heart out to the body and it's systems (see chapter 7). The surgically implanted LVAD is a battery powered mechanical pump which helps maintain cardiac output of blood.

Lobectomy

The surgical removal of part of a lung called a lobe (see chapter 7) in order to remove the tissue that is no longer working properly.

Lung Biopsy

A biopsy is obtaining representative tissue sample using special bronchoscope, needle, or surgical procedure for examination.

Lung Resection

A lung resection is partial removal (excision) of lung tissue in order to remove diseased/obstructive tissue.

Lung Transplant

A lung transplant is the grafting (replacement) of a lobe, lung, or lungs. Lobe transplant is the replacement of part of a lung and usually requires two donors. Lung transplant is the replacement of a single lung usually a living lung (from a donor). Cystic fibrosis patients who have bacterial colonization (bacteria in both lungs) require a double lung transplant (bilateral lung transplant). This surgery is required because the infection could infect a new single lung (see chapter 9).

Note: When having a heart and lung transplant the survival rate at three years is over 60%.

Mannitol Bronchial Challenge

Spirometry tests (inhaled and exhaled volumes vs. time) are done to establish normal patient values or a baseline. Mannitol (is a sugar alcohol used as a diuretic or water pill) is the administered in small amounts through a dry power inhaler to induce a 20% decrease in the established spirometry volume.

Note: This test is awaiting FDA (Food and Drug Administration) approval.

Methacholine Challenge Test (MCT)

The methacholine challenge test ordered to assess airway responsiveness. This test is useful when asthma is suspected and traditional testing methods have not supported a diagnosis. The patient completes a Pulmonary Function Test (PFT), inhales methacholine using a nebulizer and completes a second Pulmonary Function Test (PFT). The test yields a significant change in asthma patients. in order to quantify airway response. There are many contraindications (reasons you can not take the test) and the test has many limitations. If your healthcare provider prescribes the test he/she will discuss the necessity and risks involved with testing.

Minimally Invasive Heart Surgery (Port Access Coronary Bypass Surgery)

Similar to coronary bypass surgery however, the breastbone is not opened. Small incisions are made in the chest in order for instruments to be passed, and grafts are then put in position. The surgeon uses cameras and video monitors to view his operation. In most cases the heart is stopped and the patient is placed on a heart lung machine (circulates and cleans the blood while the heart is stopped). The graft result is the same as bypass surgery (see Coronary Bypass Surgery).

MRA (Magnetic Resonance Angiogram)

A Magnetic Resonance Angiogram (MRA) is a type of MRI (see below) which uses a magnetic field and pulsed radio waves to view blood vessels. It allows the doctor to see the blood vessel walls and can locate clots, bulges (aneurysms), fat or calcium buildups (plaque) narrowing the vessels (stenosis) or leaking or tearing of a vessel (dissection) in any part of the body that your healthcare provider wants to study.

MRI (Magnetic Resonance Imaging)

The MRI should be called "the big noisy beast". MRI is a test that uses a strong magnetic field and pulses of radio wave energy (like your local radio station sends out) to obtain images of organs and structures inside the body (such as heart, lungs, blood vessels etc.) often using a contrast material (dye) to get better images. The MRI provides information that cannot be seen on X-ray, ultrasound or CT scan. This makes the MRI very useful in finding problems within internal organs, locating and/or diagnosing: tumors, cancer, bleeding, aneurysms (localized over inflation of a blood vessel), injury, stroke, blood vessel problems, nerve problems, spinal problems, blockages, infection and many others.

Pacemaker

When the normal pacemaker for the heart (the SA Node, see chapter 7 & 11) is not working correctly (i.e. too fast, too slow or not regular) a pacemaker may be the ticket. The pacemaker is placed under the skin and attached to the heart through leads (wires) or can be attached directly to the heart. The impulses from the pacemaker help control the heartbeat. There are two types of pacemakers, demand (as needed) or continuous (which pace the heart or control the heartbeat).

New technologies in pacemaker models allow them to alert your doctor to early heart failure by detecting fluid buildup in your lungs up to 3 weeks before you experience a heart attack.

Pericardiocentesis

This procedure is used to remove unwanted fluid from between the heart and the sac (pericardium) caused by cardiac tapenade or cardiac effusion. A needle (with syringe) or a catheter (chest tube) is inserted through the chest wall (just above or just below the heart) into the heart sac (pericardium). The fluid is removed which allows the heart to pump more efficiently without the pressure from the excess fluid restricting its movement.

Pleurodesis

The artificial adhesion between parietal pleura (tissue covering the inner lining of the chest wall) and visceral pleura (tissue covering the lungs, see chapter 7) by surgical or chemical means. This procedure is used to correct severe or reoccurring hemothorax (blood or fluid between the lungs and chest wall) or pneumothorax (air between the lungs and chest wall). If a chest tube (drain tube placed in the chest wall, see chapter 6) is needed to apply the chemicals (tetracycline, iodine or other medications) can be introduced to the pleural space unless a surgical procedure is needed. The chemicals are irritants and cause the pleura's to bond back together. Surgical procedures include thoracotomy (surgical opening the chest to view the pleura) and thoracoscopy (examination of the pleura using an endoscope) while physically roughing up the pleura (using a rough surgical pad) to facilitate adhesion. Patients are premeditated with sedatives (medications to relax the patient) and analgesics (for pain relief), due to the fact that both procedures are very painful.

Plethysmography (see Body Plethysmography)

Pneumonectomy

A Pneumonectomy is the surgical removal of a whole lung. Pneumonectomy is required to remove cancer, hemoptysis (bleeding from the lung) or bronchiectasis when other forms of treatment have failed. Reducing the size of the lungs will reduce the total lung volumes (see PFT in chapter 11). An incentive Spirometer will be ordered after surgery to improve the breathing function and promote good lung health (see chapter 6).

Pulmonary Function Tests (PFT)

One of the main purposes of the Pulmonary Function Test (PFT) is to determine why you have dyspnea (difficulty breathing). There are two types of dyspnea Physiologic (physical exertion) and Pulmonary (Restrictive where the lungs cannot move properly and Obstructive where there is resistance to air flow).

PFT's determine how well the lungs are working and to make diagnoses between obstructive disease pattern (decreased flow rates, increased Residual Volume and increased Total Lung Capacity) and restrictive disease patterns (decreased volumes and decreased Total Lung Capacity). Obstructive diseases include asthma, bronchitis, Bronchiectasis, emphysema, cystic fibrosis and bronchopulmonary dysplasia, while restrictive diseases make up most of the rest.

Normal values are based on normal adults at their Ideal Body Weight (IBW), height, weight, sex and race. It is normal for values to decrease with age because our lungs become less efficient.

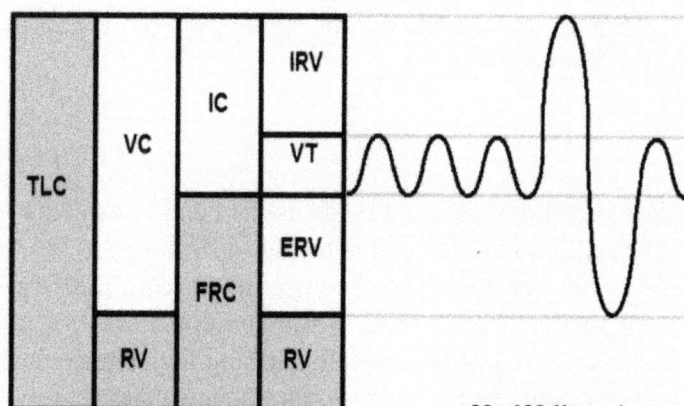

Pulmonary Function

80 - 100 Normal
60 - 79 Mild
40 - 59 Moderate
<40 Severe

PFT

Obstructive diseases (asthma, emphysema, chronic bronchitis & small airway disease) are indicated by the gray blocks. Restrictive diseases are ones that will not allow your lungs to move efficiently (the clear blocks above). Diseases are discussed in detail in chapter 9.

VT or Tidal Volume is the volume of air that is inhaled and exhaled during normal breathing.

IRV or Inspiratory Reserve Volume is the maximum volume of air than can be inhaled after a normal VT inspiration (like a yawn).

ERV or Expiratory Reserve Volume is the maximum volume of air than can be exhaled after a normal VT exhalation.

RV or Residual Volume is the volume of air remaining in the lungs after a maximal exhalation (air that cannot be removed from the lungs during breathing).

TLC or Total Lung Capacity is the maximum amount of air your lungs can hold naturally.

VC or Vital Capacity is the maximum inspiration and expiration. This function can be calculated bus is used to test for air trapping.

IC or Inspiratory Capacity is a calculation of the amount of air that can be inspired after a normal VT exhalation.

FRC or Functional Residual Capacity is a calculation of the air remaining in the lungs after a normal VT exhalation.

The PFT also measures the rates of airflow in and out of the lungs (peak flow tests the small airways for asthma), and positive/negative pressures to aid in diagnosis. However, we will not go any deeper into technical explanations.

Pulse Oximetry

Oximeters record blood oxygen levels (SpO2), pulse and the perfusion index (how well the blood is supplying oxygen to tissue). See chapter 6.

Radiofrequency Ablation

The SA Node (see chapter 7) normally has a single cell that controls the beating of your heart or pacemaker. This surgery is used when more than one cell is trying to stimulate a heartbeat. Similar to angioplasty using a catheter guided by fluoroscopy (a real time x-ray viewed on a monitor) and equipped with an electrode in order to deliver a painless radiofrequency pulse (similar to a microwave) to destroy the cells causing the abnormal heartbeat. This is the preferred method of treatment for many types of abnormal rapid heartbeats (arrhythmias) and rarely requires a repeat procedure.

Note: Similar ablations are done with laser or heat.

Sleep Study (Polysomnography)

This is a very important diagnostic tool. Not all sleep problems are related to apnea (stopping breathing) or require treatment by wearing a mask at night. Often poor sleep quality can be caused by a combination of factors such as restless legs, nocturnal asthma, obstructive sleep apnea or central sleep apnea. If you wake up at night, there is a reason; your body may be telling you that something is wrong. Your doctor may talk about poor sleep hygiene, which when put in simple terms, could include television, radio, lights, alcohol, or other things that may wake you during the night. He/she will ask about your spouse's snoring and if they wake you at night because of your snoring.

There four types of sleep studies Polysomnogram (PSG):

Diagnostic Overnight PSG is used to do general monitoring and evaluation.

Split Night PSG with CPAP or BiPAP Titration (see chapter 6) is conducted when moderate or severe Sleep Apnea is discovered during the first part of your Diagnostic Overnight PSG.

Two Night PSG with CPAP Titration (second study), (see chapter 6) is ordered when your Diagnostic Overnight PSG requires the doctor review some sleep apnea or when you sleep apnea started late in the sleep period. The second night is used for titration of CPAP or BiPAP in order to relieve your sleep apnea.

Diagnostic Daytime Sleep Latency Test (MSLT) is used to diagnose Narcolepsy (uncontrolled falling asleep during your waking hours) and in measuring daytime sleepiness. In order to ensure accuracy of the test results this test is performed on the morning after your Diagnostic Overnight test. It consists of five (5) naps scheduled two hours apart during the day.

While no one likes to sleep somewhere else (unless on vacation) the data collected can really enhance your life, by indicating the need for corrective treatments. Due to advances in technology, some sleep centers are capable of performing sleep studies in your home. You can speak with your healthcare provider for availabilities in your area.

When you get to the sleep center with your overnight bag (i.e. toothbrush, pajamas etc.) you will be escorted to your room with a private bath. Your sleep technician will allow you to prepare for bed and then setup your monitoring system.

> The EEG (Electroencephalogram) measures and records brain wave activity (alpha, beta, delta and theta waves) in order to determine your stages of sleep.

> The EMG (Electromyogram) records muscle activity; facial movements, twitches, teeth grinding, and leg movement during stages of sleep.

> EOG (Electro-oculogram) records eye movement during stages of sleep (like REM).

> EKG (Electrocardiogram) records heart rate, electric rhythm and activities.

> Nasal/Oral Airflow Sensor records breathing rate, airflow and breath temperature.

> Chest/Abdominal Belts record the depth of your breaths.

> Oximeters record blood oxygen levels (SpO2) and some have capnometer that measures carbon dioxide (CO_2).

> Snore Microphone records snoring.

> Video records your body position and movements (only the doctor will review the video tape).

Note: None of these monitoring devices hurt or have anything sharp associated with them.

During your Diagnostic Overnight PSG if obvious signs of obstructive sleep apnea are present early in your study the Split Night PSG will begin and a CPAP/BiPAP will be titrated in order to relieve your apnea symptoms. This does not hurt either and you may wake with the best night's sleep you have had in years.

If someone has told you "you snore" or "you stop breathing while you sleep", please tell your healthcare provider and take the time to be tested. Serious problems are diagnosed during the sleep study and your quality of life could be greatly enhanced by identifying and correcting problems with your sleep pattern. These tests can save or prolong your life.

Note: Once you have had a sleep study it is not usually necessary to repeat it unless you original symptoms return (this is rare).

Spirometry

Spirometry is the physical measurement of how much you inhale and exhale during specific time period.

The specific values that spirometry measures are Forced Expiratory Vital Capacity (or FVC), the Forced Expired Volume in one second (or FEV1) and the Forced Expiratory Flow Rate 25-75. Normal patient values are based on similar populations (sex, race, age, height, weight etc.).

Note: The names are not important for you to remember.

Stent Procedure

A small wire mesh tube (looks like a small ball point pen spring) is placed at the plaques location within the coronary artery (blood supply to the heart muscle) and is released. The mesh tubes (stents) open expanding the arteries and are left in place in order to restore oxygenated blood flow to the heart. First performed in 1987.

Tilt Table Test

If you have experienced fainting with no known cause this may be your test. You lie supine (flat on your back) and the table will be tipped slightly head down then to a standing up position. Your healthcare provider will check your reaction (fainting, blood pressure, heart rate, etc.) to changing position. This assists your healthcare provider diagnose why you get dizzy when you set up or stand up (orthostatic hypotension) or that thee cause is vagus nerve (neurally mediated hypotension, see Valsalva Maneuver).

Thoracentesis (Needle Aspiration or Pleural Tap)

This is not actually a surgical procedure, it utilizes a needle inserted into the chest between the lung and chest wall in order to remove excess pleural fluids or air from between the lungs and chest walls.

Thoracotomy

This is a major surgery, where an incision is made into the chest in order to gain access to the heart, lungs or esophagus. There are three type of thoracotomy surgery; Median Sternotomy which is through the sternum (like most open heat surgeries through the center of the chest), Posterolateral Thoracotomy through the back of the chest (like most lung surgeries) and Anterolateral Thoracotomy through the front of the chest on the left side (used in traumatic cardiac arrest).

Tracheostomy

This surgical procedure opens a semi-permanent or permanent opening without using a tracheostomy tube (trach tube, see chapter 6). An opening is made midline (center) of the neck between the sternal notch (the "U" shaped bone between the chest and base of the neck) and the cricoid cartilage (the Adam's apple or bump on the front of the throat, which is larger on men). The permanent or semi permanent opening in the neck (called a stoma) which is used for breathing. Humidification may be required in order to maintain lung tissue health (the nose would normally provide the needed humidification).

Tracheotomy

The procedure is used by paramedics and doctors to establish an alternate airway to the lungs when the mouth and nose can not be used. After the incision is made midline (center) of the neck between the sternal notch (the "U" shaped bone between the chest and base of the neck) and the cricoid cartilage (the Adam's apple or bump on the front of the throat, larger on men) the tracheostomy tube (trach tube, see chapter 6) is inserted to secure the airway. These trach tubes may or may not have a balloon at the tip to secure the airway.

Transmyocardial Revascularization (TMR)

The initial incision is made on the left breast in order to expose the heart. A laser drill is used to make a series of holes into the hearts chambers. In rare cases this procedure is combined with regular bypass surgery. This procedure is normally used in patients who are not candidates for coronary bypass surgery.

Treadmill Exercise Challenge

Similar to a cardiac stress test, this test is ordered when a patient complains of difficulty breathing (dyspnea) during exercise. The patient walks or runs on the treadmill until their heart rate is 80-90% of the predicated value and continues for 3-4 minutes. After exercise spirometry is measured at 1 minute, 5 minutes, 10 minutes and 20 minutes for a decrease (10% to 15%) of the Forced Expired Volume in one second (or FEV1, see spirometry).

Note: This test may be completed during a Cardiac Stress Test.

Uvulopalatopharyngoplasty

This surgical procedure used in treating severe obstructive sleep apnea. It consists of shortening the soft palate and removal of the uvula and tonsils

Valsalva Maneuver

This is a fast and simple test designed to test your nervous system associates with your heartbeat and blood vessels. Your healthcare provider will ask you to take a deep breath and blow (as hard as you can) through pursed lips. You will be asked to repeat the test several times. Your heart rate and blood pressure will checked several times for changes.

V/Q Scan or Ventilation Perfusion Scans

Measures gas and blood flow distribution (ventilation & perfusion). The patient inhales a small amount of radio labeled gas (xenon), and receives an injection of radioisotope (type of dye), followed by study of the mismatches between ventilation & perfusion. The test detects poorly ventilated areas of the lungs (such as a pulmonary embolism etc) and reduced perfusion (blood flow) of the small blood vessels.

Wedge Resection (Segmentectomy)

The removal of a smaller portion or segment of a lung (see chapter 7).

X-Rays

X-rays utilize ionizing radiation in order to see inside the body. The chest x-ray displays the condition of the trachea, right/left main stem bronchi, all the lobes of each lung, the diaphragm, vessels within the lungs, the heart, aortic knob (notch above the heart), veins and arteries. In addition, abnormal things such as solids, gases or fluids located in an area they do not belong are easily seen.

Note: Air (pneumothorax) or fluids (hemothorax) between the chest wall and lung will make breathing difficult. Extra fluids around the heart (cardiac tamponade) make it difficult for the heart to pump blood efficiently. These problems were discussed in chapter 9.

X-rays are generally named for how the image is taken (i.e. front to back, left to right etc.). You would be amazed to know how many types of x-ray positions actually exist. When ordering an x-ray your healthcare provider will consider what he/she wants to view (each type of x-ray has a specific diagnostic value) and the patient's condition or limitations. The most common x-ray taken in the hospital is the A-P or anterior-posterior (front to back). This is the typical portable chest x-ray used for the bedridden, critically ill, or difficult-to-move patients.

The x-ray picture consists of high-density (bone or solid organs) & low-density structures (air filled). High-density structures appear white (radiopaque or radio dense) on an X-ray, because more of the x-ray energy is absorbed, and less of the x-ray reaches the film. The higher the density, the whiter it appears. The bones are the densest naturally occurring structures, followed by fat, then water. Low-density structures appear dark (radiolucent) on x-ray because less of the x-ray is absorbed and more of the x-ray reaches the film. Normal lungs appear radiolucent because they contain air, which has the least density of all. Bones such as the clavicles, ribs, or sternum are easily located and make good landmarks for locating other structures or organs.

Notes:

Chapter 12

Conversion Charts

	Conversion Formula		Examples		
Fahrenheit	to	**Celsius**	F	C	
	-32*5/9 =		0	-17.78	
	5/9 = 0.555556		32	0	
			98.6	37	
Celsius	to	**Fahrenheit**	C	F	
	*9/5 + 32 =		0	32	
	9/5 = 1.8		32	89.6	
			100	212	
Ounces	to	**Milliliters**	Oz	ml	
	*30 =		1/6	5	1 tsp
			1	30	2 tbsp
Milliliters	to	**Ounces**			
	*0.034 =				
			1 Liter	1000 ml	
Pounds	to	**Kilograms**	lbs	kg	
	*0.45 =		120	54	
			150	68	
Kilograms	to	**Pounds**	kg	lbs	
	*2.2 =		50	110	
			100	220	
Inches	to	**Centimeters**	in	cm	
	*2.54 =		60	152.4	
			72	182.9	
Centimeters	to	**Inches**	cm	in	
	*0.4 =		142.24	56	
			172.72	68	
	* = multiply				

Note: Formulas are approximant and are intended as guidelines.
Note: 1 cc is the same as 1 ml

The International Unit (IU) is a pharmacological measurement based on the amount of a substance required for an effect or activity. The measurement includes medications, vaccines, vitamins, hormones and others. For instance; one IU of vitamin C (50 micrograms) does not contain the same number of micrograms as one IU of vitamin D (0.025 micrograms).

Glossary

Medical conditions and diseases have a unique language. In order to research and understand medical definitions, causes, symptoms, tests and treatment options you may need a little help understanding that language. This is in no way a complete listing of terms you may uncover however, it is a start. You may want to purchase a book on basic medical terminology to help you in your research outside of this book.

Ablation	destruction or removal of tissue
Abscess	puss collection within tissue, organs or in any confined space inside the body caused by a bacterial infection
Acidemia	arterial blood is more acidic than normal (pH < 7.35)
Acidosis	a physiological process resulting in an increase accumulation of an acid or the loss of base
Acute	severe or with sudden onset and short time span
Aerobic exercise	physical activity that requires increased cardiac output and ventilation to meet the increased oxygen demands
Aerosol	a solution of a drug that is made into a fine mist for inhaling
Airborne precautions	safeguards designed to reduce the risk of airborne transmission of infectious organisms by droplets or dust
Airway	breathing tubes that carry air into and out of the lungs
Airway obstruction	a narrowing or blocking of the passages that carry air to the lungs
Alkalemia	arterial blood is less acidic than normal (pH > 7.45)
Alkalosis	a physiological process resulting in an increase accumulation of a base or the loss of acid
Alveoli	tiny air sacs at the end of the airways
Anaphylaxis	an exaggerated hypersensitivity reaction
Anemia	low red blood cell count, reduces the body's ability to carry oxygen
Aneurysm	localized swelling of the wall of a blood vessel
Angina pectoris	severe chest pain caused by reduced blood flow to the heart, often radiates to the shoulders and/or arms
Angiogenesis	drug induced or spontaneous growth of new blood vessels
Angiography	an x-ray after administering a radiopaque contrast showing the internal anatomy of the heart and blood vessels
Antiarrhythmic	medication used to treat abnormal heart rhythms
Antioxidant	Vitamins A, C and E which may limit cellular damage caused by free radicals
Anti-inflammatory	medications that reduce inflammation
Anti-rejection Meds	medications that help to prevent the body from attacking a new organ
Aortic Valve Homograft	replacing the aortic valve from a human donor
Aphasia	neurological condition where language function is defective or absent
Apices	top part of the lungs
Apnea	a stoppage of spontaneous breathing apnea - hypopnea
Arrhythmia	irregular heart beat or rhythm
Arteriole	one of the smallest branch arterial blood vessels
Artery	a blood vessel which carries oxygenated blood from the heart to the rest of the body

Asthma	respiratory condition caused by episodic narrowing and inflammation of the airways in response to a trigger; symptoms include wheezing, coughing, shortness of breath, and labored breathing
Aspirate	remove a fluid through suction
Aspiration	inhaling food, vomit or foreign material into the lungs or the process of withdrawing fluid through suction
Asymptomatic	without showing any symptoms
Atomizer	a device that produces an aerosol
Atrial Fibrillation	a very rapid beating of the atria which does not allow the atria to fill with blood and preload the ventricles
Atrial Flutter	a rapid beating of the atria which does not allow the atria to fill with blood and preload the ventricles
Atrium	top of both sides of the heart that preloads the ventricles (like taking a deep breath before blowing up a balloon)
Atrophy	the wasting muscle tissue due to disease or other influences
Bacteremia	bacteria circulating within the blood
Barrel chest	the abnormal increase in the front to back diameter of the chest due to over inflation of the lungs
Base	like baking soda it reduces acidity in the blood
Biot's Respiration	abnormal breathing pattern characterized by irregular periods of apnea alternating with periods of four or five breaths of identical depth
BiPAP	Bi-Level Positive Airway Pressure or continuous positive airway pressure with inhalation pressure as well as a lower exhale pressure for comfort
Blood Brain Barrier	the feature of the brain where the walls of capillaries in the central nervous system and surrounding membranes separates the central nervous system from your blood. It slows or prevents some medications, chemicals, compounds and disease causing organisms from reaching the brain or central nervous system
Bradycardia	slow heart rate (<60 beats per minute)
Bradypnea	Slow rate of breathing
Bronchi	the larger branching airways dividing into the lungs' lobes and segments
Bronchial hygiene	the clearance of mucus using Chest Physical Therapy (CPT), vibratory vest therapy, producing an effective cough, Bronchoscopy or by using equipment discussed in chapter 6
Bronchioles	smaller branching airways about 2 mm in diameter or less
Bronchoconstriction	bronchi narrowing due to contraction of smooth muscle
Bronchodilation	the reversal of bronchoconstriction, usually via sympathetic stimulation
Bronchography	an x-ray examination of the bronchi after they have been coated with a radiopaque substance
Bronchophony	abnormal voice sounds heard over sick (consolidated) lungs
Bronchorrhea	an excessive discharge of respiratory tract secretions
Bronchospasm	the constriction of smooth muscle of the bronchi resulting in narrowing and/or obstruction
Bruits	an abnormal sound heard from the heart or large vessels, caused by turbulence (like folding a garden hose)

Bundle Branch Block	normally electrical impulses travel down the right and left bundle branchs at the same time inorder to have the right and left ventricle beat at the same time; a small blockage in one branch desynchronizes the heart beat of the ventricles (see chapter 8 & 11, Rabbet Ears)
Capillaries	the smallest of the blood vessels, only 1 cell thick
Capnography	the process of obtaining of the amount of carbon dioxide in expired air using a capnograph or capnometer indicating the quality of lung function
Capnometer	an instrument used in anesthesia, respiratory physiology, and respiratory care to measure the proportion of carbon dioxide (CO_2) in expired air indicating the quality of lung function
Capnograph	the measurement of carbon dioxide (CO_2) in a volume of gas, usually by methods of infrared absorption or mass spectrometry
Carbon Monoxide (CO)	Colorless and odorless gas produced by flame or burning
Carboxyhemoglobin	a compound produced by the chemical combination of hemoglobin with carbon monoxide (CO)
Carboxyhemoglobinemia	a decrease in the oxygen carrying capacity of the blood due to the saturation of hemoglobin with carbon monoxide (CO)
Cardiac tamponade	compression of the heart due to the collection of blood, fluid, or air in the pericardium
Cardiomegaly	enlarged muscle tissue (hypertrophy) of the heart usually caused by pulmonary hypertension
Cardiomyopathy	any disease that affects the heart (myocardium)
Cardioversion	the restoration of the heart's normal sinus rhythm by electric shock to the patient's chest
Carina	the point of separation of the right and left mainstem bronchi
Cartilage	non-rigid connective tissue
Cerebrovascular	the vascular system of the brain
CFC	chlorofluorocarbons propellant used in Metered Dose Inhalers (MDI)
Chemoreceptor	a sensory nerve cell which stimulates breathing due to changes oxygen or carbon dioxide (CO_2) levels in the blood
Clot	anything that can get stuck in a vein or artery (i.e. coagulated blood, tissue, foreign body etc.)
Congenital	existing from birth
Consolidation	a process of becoming solid; especially applies to lung tissue due to fluid collection (like in pneumonia)
Contact precautions	safeguards designed to reduce the risk of transmission of disease
Costophrenic Angles	where the bottom of the lung on the outside meet the ribs
Crepitus	a dry crackling sound which is felt when touching (palpating) an area of subcutaneous emphysema
Croup	an infection of the upper airway occurring in infants and children
Cyanosis	an abnormal bluish discoloration of the skin
Deadspace	air which does not experience gas exchange within the lungs
Defibrillation	an electric counter shock to the heart delivered by a defibrillator
Desaturation	a significant drop in oxygen level in the blood
Dialysis	a process of removing crystalline and colloids instead of using the kidneys

Diaphragm	the large dome-shaped muscle or the primary muscle of ventilation
Diastolic blood pressure	the bottom or lower number which indicated the pressure of arterial blood when the heart is resting
Diminished	reduced sounds due to a reduction in flow
Droplet precautions	safeguards designed to reduce the risk of transmission of disease through droplet transmission
Drug Interactions	any foods, prescribed medications, over the counter medications/herbs/supplements, or drinks that alter the effect of the prescribed medication
Ductus Arteriosus	a fetal vascular bypass connecting the pulmonary artery directly to the descending aorta (normally closes after birth)
Dysphagia	difficulty in swallowing
Dysphasia/	impaired speech
Dysphonia	paralyzed vocal cords
Dysoxia	an abnormal metabolic condition where the body is unable to properly use the oxygen
Dyspnea	Shortness of breath; difficult or labored breathing
ECMO	abbreviation for extracorporeal membrane oxygenation; where venous blood is pumped outside the body to a heart-lung machine for oxygenation and returned to the body
Edema	excess fluid buildup
Effusion	due to blood vessel rupture fluid collects into a body cavity
Egophony	the normal sound voice heard through the chest wall
Embolism/Embolus	a blood clot, tissue, tumor, air, fat or substance lodges in a small blood vessel reducing flow or causing a complete blockage
Endocrine system	the network of glands and structure that secrete hormones (targeting specific organs) directly into the bloodstream
Endotracheal	inside the trachea
Epiglottis	a cartilage extending from the base of the tongue backward
Epistaxis	bleeding from the nose
Etiology	the study of the cause
Exacerbate	to worsen
Exacerbation	a worsening of a condition normally acutely (quickly)
Extubate	removing the ET tube from the trachea
Exudates	a high protein fluid that escapes out of a cell (caused by inflammation or infection)
Febrile	a fever
Flaccid	weak or lacking normal muscle tone
Flail chest	a traumatic multiple site chest injury where rib cage becomes unstable
Flowmeter	a device that controls the flow rate of a medical gas
Free Radical	destructive fragment of oxygen which are believed to trigger atherosclerosis (plaque build up in arteries)
Fremitus	a vibration of the chest wall that can be heard (auscultated) or felt (palpated)
Fungal	yeast, molds and mushroom, separate from plants, animals or bacteria
Fungicide	an agent which kills fungi

Gas exchange	primary function of the lungs; transfer of oxygen from inhaled air to the rest of the body and of carbon dioxide (CO2) from the rest of the body into the lungs for exhalation
Gram Negative	a bacterial disease of the family "coca' as in staph infections
Gram Positive	a bacterial disease of the family "bacilli" such as anthrax or salmonella
Heart Block	an arrhythmia caused by a partial or complete blockage of electrical pathways of the heart (see chapter 7 & 11)
Heliox	a therapeutic mixture of helium and oxygen used to treat large airway obstructions
Hemorrhage	bleeding
Hemodialysis	a procedure where impurities or wastes are removed from the blood (see dialysis)
Hemoptysis	coughing up blood
Hemothorax	an accumulation of blood and fluid in the pleural space, between the chest wall and lungs
HFC	hydroflouroalkane propellant used in Metered Dose Inhalers (MDI)
Huff	a type of forced cough or forced expiration to replace coughing when pain limits normal coughing
Humidifier	a device that adds molecular water to humidify a gas
Hyperbaric oxygen	therapy in which oxygen is delivered at a pressure greater than 1 atmosphere (like a pressure chamber for divers)
Hypercapnia	presence of excess amounts of carbon dioxide (CO2) in the blood
Hypercapnic	the inability to maintain the normal carbon dioxide during levels during respiratory failure
Hyperchloremia	presence of excess amounts chloride in the blood
Hyperkalemia	presence of excess amounts of potassium in the blood
Hyperlucent	in an X-ray; a low density area which appears clear or transparent (boney areas would appear white)
Hypermagnesemia	presence of excess amounts of magnesium in the blood
Hypernatremia	presence of excess amounts sodium in the blood
Hyperphosphatemia	presence of excess amounts phosphate ions in the blood
Hypersensitivity	an undesired reaction by the immune system, which requires a pre-sensitized state of the immune system of the patient
Hypertension	an abnormal high blood pressure
Hypertrophy	the increase in size of a tissue or organ
Hyperventilation	rapid deep breathing (exceeding ventilation needs > 20/minute)
Hypervolemia	presence of excess amount fluid circulating blood
Hypocalcemia	presence of decreased amount of calcium in the blood
Hypocapnia	presence of decreased amount of carbon dioxide (CO2) in the blood
Hypochloremia	presence of decreased amount of chloride level in the blood
Hypoglycemia	presence of decreased amount of glucose (sugar) in the blood
Hypokalemia	presence of decreased amount of potassium in the blood
Hypomagnesia	presence of decreased amount of magnesium in the blood
Hyponatremia	presence of decreased amount of sodium in the blood
Hypophosphatemia	presence of decreased amount of phosphate in the blood
Hypopnea	shallow breathing
Hypotension	an abnormal low blood pressure

Hypothermia	an abnormally low body temperature
Hypoventilation	ventilation less than that necessary to oxygenate the blood
Hypovolemia	an abnormally low blood volume
Hypoxia	an abnormally low oxygen level in the blood
Idiopathic	a disease or process with no known cause
Infarction	localized tissue death (necrosis)
Infectious	any foreign species capable of colonization (growth or reproduction) of itself in the host
Infiltrate	an unwanted material within the lungs
Informed consent	the principle of law that states that a patient must be informed and understand a procedure prior to proceeding
Insomnia	inability to sleep
Intercostal	pertaining to the space between two ribs muscle
Interstitial	fluid outside of the vascular spaces
Intubation	the insertion of an endotracheal tube within the trachea
Laryngospasm	the involuntary contraction of the laryngeal muscles
Lobule	a small lobe section of the lung
Meconium	fetus aspiration or inhalation of meconium
Medical Advocate	an informed person who makes medical treatment decisions
Medication Interactions	any foods, prescribed medications, over the counter medications/herbs/supplements, or drinks that alter the effect of the prescribed medication
Morbidity	the state of being ill
Muscle fatigue	condition in breathing which results in respiratory failure
Narcolepsy	a sudden sleep attack
Nasopharynx	the upper airway behind the nasal and oral cavities
Nasotracheal	the passageway from the nose to the trachea
Naturopathic Doctors	teach diet, exercise, lifestyle changes and natural therapies to enhance the body to fight disease
Nebulizer	a device that produces an aerosol
Necrosis	local tissue death
Neonatal	birth to 28 days
Neoplasm	an abnormal growth of cells (does not have to be cancer)
Nosocomial infection	an infection acquired in the hospital
Novel	meaning new in medical terms
Obstructive diseases	reduce airflow
Orotracheal	the passageway from the mouth to the trachea
Orthopnea	labored breathing in the reclining position
Orthostatic	associated with standing upright with hypotension
Overlap syndrome	the combination of Obstructive Sleep Apnea (OSA) and Chronic Obstructive Pulmonary Disease (COPD)
Oximeter	a photoelectric device used to asses oxygen saturation
Oximetry	the process of using oximeters, to determine blood oxygen
Ozone	a light blue gas with 3 oxygen molecules, considered to be therapeutic in some countries
Pallor	paleness or absence of color in the skin
Palpitation	a pounding, fluttering or racing of the heartbeat
Paradoxical breathing	abnormal breathing pattern associated with a flail chest
Parasite	an organism that requires a host to live on or in
Pectoriloquy	the transmission speech sounds through the chest wall

Percussion	vibrating the chest and back in order to help move secretions
Pericardium	a fibrous sac surrounding the heart
Plethysmograph	a human size booth used to measure pressures in pulmonary physiology
Pleura	a thin layer covering the lungs and lines the inside of the thoracic cavity (chest)
Pleural effusion	the abnormal collection of fluid in the pleural space
Pleural Empyema	a pleural effusion where the fluid is purulent or contains pyogenic organisms
Pleural space	the space between the visceral pleura (lung covering) and the parietal pleura (chest wall cover)
Pneumothorax	the presence of air in the pleural space of the
Postural Drainage	positioning a patient in order to remove excess fluids from a targeted portion of the lungs
Progressive	Increasing in severity over time
Pulmonary edema	presence of excess amount of fluids within the lung
Pulmonary embolism	the blockage of a pulmonary artery
Pulmonary infiltrate	an unwanted material within the lungs
Pulmonary surfactant	a detergent-like substance that lubricates the alveoli
Pulse pressure	the difference between systolic and diastolic blood pressure
Radio aerosol	a radioactive isotope aerosol
Radiolucent	a tissue that readily permits the passage of x-rays
Radiopaque	a tissue that does not readily permit the passage of x-rays (appears white on an x-ray)
Radon	colorless, odorless and tasteless radioactive gas (element # 86 Rn) which is much heavier that air
Rales	lung sounds of the chest crackle is now preferred
Reconditioning	physical activity to strengthen essential muscle groups
Respiratory failure	a condition where the exchange of O2 and CO2 between the alveoli and the pulmonary capillaries is inadequate
Respiratory insufficiency	a condition where breathing has abnormally signs or symptoms
Restrictive diseases	restrict lung movement
Retractions	the sinking inward of the skin around the chest cage with each inspiration
Rhonchi	abnormal sounds heard in a respiratory airway obstructed by thick secretions, muscular spasm or neoplasm
Rhonchial Fremitus	abnormal vibration felt on the chest wall due to narrowed airways
Septic Shock	when infection prevents oxygen from being supplied to the cells Shock an inadequate or lack of blood supply to vital organs
Sonography	an imaging of deep structures of the body using high-frequency sound waves
Sphygmomanometer	a blood pressure cuff and pressure gauge
Spirometer	measures lung volumes or flows
Stenosis	narrowing
Stridor	a high-pitched barking sound
Stroke volume	volume of blood ejected by the left ventricle of the heart
Sublingual	beneath the tongue
Supplemental Oxygen	extra oxygen supplied as required to maintain the correct oxygen

	levels within arterial blood
Symptom	any indication of disease noticed or felt by a patient; in contrast, a sign of an illness is an objective observation
Symptomatic treatment	therapy that eases symptoms without addressing the cause of disease or symptoms
Systolic blood pressure	the top or higher number which indicated the pressure of arterial blood when the heart is contracting or pumping
Tachycardia	an abnormally fast heart rate (>100)
Tachypnea	an abnormally fast breathing rate (>20)
Tension pneumothorax	air in the pleural space that exceeds atmospheric pressure
Transcutaneous	across the skin
Trendelenburg position	a position in which the head is low and the body and legs are on an inclined
Turbinate	bony structures of the interior nasal passages
Turgor	the normal resiliency of the
Tumor	any abnormal growth of cells (does not have to be cancer)
Ultrasonography	the use of an ultrasound to see internal organs and structures
Uvulopalatopharyngoplasty	a surgical procedure used in treating severe obstructive sleep apnea. It consists of shortening of the soft palate and removal of the uvula and tonsils
Vagus Nerve	nerves that provide autonomic movement of the heart & lungs
Vasoconstriction	narrowing of the blood vessels
Vasodilation	widening or distention of blood vessels
Venule	the small blood vessels that gather blood from the capillaries
Ventricular Fibrillation	also called V-fib it is a rapid erratic quivering of the ventricles (this is a life-threatening rhythm)
Ventricular Tachycardia	also called V-tach it is a rapid beating of the ventricles without pumping blood (this is a life-threatening rhythm)
Virus	the smallest parasites (microscopic) who are dependant on living cells
Wheezes	a continuous or intermittent lung sound
Wolf Parkinson White	also called WPW is it an abnormal electrical conduction pathway that will short-circuit around the heart (can cause supraventricular Tachycardia or a very fast heart rate originating above the ventricles)
Xiphoid	part of the sternum
Xiphoid Process	the bottom pointed part of the sternum

Notes:

Index

213

COPD vii, 1, 12, 22, 49, 60, 63, **119**, 120, 126, 148, **201**

Coronary Angiogram **187**

Coronary Artery Calcification **188**

Coronary Bypass 57, 123, 125, 147, 166, **186, 196**, 202

Corona Virus **100, 162**

Corticosteroid 7, 8, **178-179**

Costophrenic Angle 153, 169, **207**

Cough Assist **70**

CPAP vii, 36, 47, 50, **60-61**

CPR **23-26**, 89, 184

Crackles & Rales **95**

Cricothyroid **82,-84, 188**

Cricothyrotomy **188**

Cricoid Cartilage **78**, 83-84, **201**, 202

CT Scan **106**

Cyanosis 93, **207**

Cylinder Oxygen 32, 33, **52**, 57

Death 1, 20, 31, 36, 37, 38, **44**, 92, 186, 195

Deep Vein Thrombosis 20, 120, **127**, 158

Defibrillator **24**, 184, & **194, 207**

Dehydration **93**, 178

Dementia 32, 37, 164

Dentures **36**, 61

Desaturation **55**, 63, **207**

Deviated Septum **79**

Diagnose vii, 3-4, 16, 92, 187, 197, 199, 200, 201

Diastolic Blood Pressure **93-94**, 128, 138, 140, 192, **208**

Diminished **95, 208**

Discharge Plans **12**

Disease Categories **101**

Dissection **196**

Diuretics 1, 7, 105, **180-181**, 195

Disinfection 33, **50**

Donor 38, **42**, 193, 195, 205

DPI vii, 66, **68**-69, 70, 176-178

Drug Interactions
Alcohol, Caffeine, Candy, Chocolate, Citrus, Dairy, Fermented Foods, Fiber, Garlic, Green Veggies, Nuts, OTC Medications, Soy, Tobacco, **6-8, 208, 210**

Dry Powder Inhalers vii, 66, **68**-69, 70, 176-178

Dust Pneumonia 116, **129**

DVT 20, 120, **127**, 158

Dying 1, 20, 31, 36, 37, 38, **44**, 92, 186, 195

Dyspnea 12-13, 18, 19, 20, 35, 49, 50, 89, 197, 202, **208**

E-cigarettes **21**

Echocardiography or Echo **187**

ECMO **72**, 105, 122, **208**

Edema 193, 96, 179, 180, **208, 211**

EEG 164, **192**, 200

Ejection Fraction 123, 128, 136, **189**

EKG vii, 19, 81, 89, 92, **189**, 193, 194, 200

EKG 12 Lead **190-192**

EKG 3 Lead **189**

Electrical Power Loss **33**

Electrocardiogram vii, 19, 81, 89, 92, **189**, 193, 194, 200

Electroencephalogram 164, **192**, 200

Electromyogram **192**, 200

Electronic Cigarettes **21**

Emergency List 30, **43**, 174

Electronic Medical Record **19**

Emergency Medications **30, 174**, 176

EMG **192, 200**

Endotracheal Tube **70**, 176, 210

Endobronchial Ultrasound **193**

Electro-oculogram **192, 200**

EOG **192, 200**

Epiglottis **78-79**, 83, 85, 131, **208**

Equipment Cleaning 33, **50**

Equipment Troubleshooting **22-24**, 60, 63

ERV **198**

Esophagus **78-79**, 198

ET Tube **70**, 176, 210

Eustachian Tube **78-79**

Event Monitor **193**

Exacerbation **208**

Exercise **12-16**, 18, 21, 33, 35, 49-50, 56, 92, 188-189, 194, 202, **205**

Exhaled Nitric Oxide Analysis 110, 172, **193**

Expiratory Reserve Volume **198**

Exsufflator Insufflator **70**

Facilities 2, 19, 34

FAA 25, 52

Fainting 124, 166, **201**

215

217

218

BOLD page numbers indicate the
main page or the explanation.

&   Between the page numbers
indicates the index item has
more than one meaning.

www.ingramcontent.com/pod-product-compliance
Lightning Source LLC
Chambersburg PA
CBHW081407200326
41518CB00013B/2268